Meaning Making in Planning

Planning theorists normally focus on issues of contest and critique. The field of planning theory is thereby replete with studies of conflict, collaboration and criticism. Considerably less critical attention is afforded to policy approaches that emerge, evolve and are widely adopted in the apparent absence of discord. This book addresses this knowledge gap.

A case study of the emergence of green infrastructure policy in Ireland is used to both inform and illustrate a theory of 'Policy Entitlement'. This interpretive approach focuses on meaning making in context to explain the counter-intuitive processes through which a new policy concept can emerge and reprofile planning activities by producing the seemingly pre-existing objective reality to which such policy is then applied and the discipline (re)orientated. This approach accounts for how a new planning concept can appear to resolve problematic policy ambiguity by suspending disagreement on issues where dispute could be expected.

This book will be of interest to those studying planning theory and the policy process, as well as those concerned with the undertheorised but swift rise to prominence of green infrastructure planning.

Mick Lennon is Associate Professor in the School of Architecture, Planning and Environmental Policy at University College Dublin, Ireland. His work focuses on the intersections between planning and environmental policy, with a particular emphasis on using novel theoretical approaches to cultivate resilience in the context of social and environmental transitions. Mick was a practicing town planner for several years prior to entering academia. He is author of *Planning for the Common Good* published by Routledge in 2022.

Meaning Making in Planning

Theorising 'Policy Entitlement' in the
Emergence of Green Infrastructure Planning
in Ireland

Mick Lennon

Routledge
Taylor & Francis Group

LONDON AND NEW YORK

First published 2024
by Routledge
4 Park Square, Milton Park, Abingdon, Oxon OX14 4RN

and by Routledge
605 Third Avenue, New York, NY 10158

Routledge is an imprint of the Taylor & Francis Group, an informa business

British Library Cataloguing-in-Publication Data
A catalogue record for this book is available from the British Library

ISBN: 978-1-032-53583-8 (hbk)
ISBN: 978-1-032-53786-3 (pbk)
ISBN: 978-1-003-41360-8 (ebk)

DOI: 10.4324/9781003413608

Typeset in Times New Roman
by Taylor & Francis Books

To Rich Lomas

Contents

List of Figures		viii
Preface		ix
	Introduction	1
1	Interpreting Policy Discourses	11
2	Case Study Summary	38
3	Narrative Production	47
4	Naming Attributes and Effects	59
5	Rationality Resonance	83
6	Narrative Modality	105
7	Policy Entitlement	125
	Index	144

Figures

4.1 Diagrammatic Representation of the Relationships between the 'Naming Effects' of GI's *Entitlement* 77

5.1 Diagrammatic Representation of the Relationships between the *Naming Effects* of GI's *Entitlement* and *Rationality Resonance* 102

6.1 *Elective Rebranding* & *Solution Adoption* in 'Narrative Volunteering' 117

6.2 *Imposed Rebranding* and *Problem Appropriation* in 'Narrative Application' 121

6.3 Narrative Volunteering and Narrative Application 122

7.1 The *Naming Effects* of Entitlement 129

7.2 The Relationships between *Naming Effects* and *Rationality Resonance* 132

7.3 The Relationships between *Naming Effects, Rationality Resonance* and *Myth* 134

7.4 Diagrammatic Representation of *Policy Entitlement* in the Relationships between *Naming Effects, Rationality Resonance, Myth* and *Modality* 136

7.5 Artificial Inflation of Discourse Coalition Size by *Narrative Application* 137

Preface

Interpretive policy analysis in planning was thin on the ground when I began researching. At that time, political sociology seemed a more fertile space for such work. While much of this work provided food for thought, it didn't seem to offer adequate keys to unlock the material I was wrestling with. But serendipity was kind to me when one day I began leafing through the pages of Dvora Yanow's *How Does a Policy Mean?* Her book offered a way of seeing the world and a language to explain it that helped confirm what I was thinking and doing. Since then, the field of interpretive policy analysis has expanded. Although still not mainstream, interpretive analysis has also grown in representation across journals, books and conferences in the subject of planning. There is now a broadening library of material available to the planning researcher setting out on their interpretive journey. I hope that this short book joins that library and proves useful to someone seeking guidance on what interpretive research in planning might look like.

This book draws on work I did over a decade ago while a PhD student at Cardiff University. I was fortunate to be supervised while there by Richard Cowell and Andrea Collins. I owe both an enormous depth of gratitude. Supporting them were Huw Thomas, Neil Harris and Peter Feindt. I sincerely thank all for their knowledge, skill and unwavering good humour. I also thank Caroline Church at Routledge for guiding me through the publishing process and the anonymous reviewers who provided feedback on a selection of chapters.

I wish to acknowledge Bristol University Press and Policy Press for permission to publish parts of this book previously published in Lennon, M. (2014). "Presentation and persuasion: the meaning of evidence in Irish green infrastructure policy", *Evidence & Policy*, 10(2), 167–186. I also wish to acknowledge SAGE Publishing for permission to publish parts of this book previously published in Lennon, M. (2015) "Explaining the currency of novel policy concepts: learning from green infrastructure planning". *Environment and Planning C: Government and Policy*, 33, 1039–1057.

Introduction

Focus and Approach of the Book

Problematic Policy Ambiguity

Policies are statements on what ought to happen. Consequently, the policy process is a site in which aspirations emerge, are debated and given representation. Scrutinising the public policy process thereby provides insight into how public interests are conceived, what these are deemed to be, and why some interests are given priority over others. In its broadest sense, this book seeks to provide such insight.

Despite decades of critique (Fischer and Forester, 1993; Yanow, 2000; Wagenaar, 2007; Bevir and Rhodes, 2015; Durnová, 2022), the policy process is conventionally conceived and practiced as applied problem-solving (Birkland, 2015; Hill and Varone, 2021). From this position, policy making is understood as a progression from problem identification to solution specification. Where difficulties arise in formulating solutions, these are seen to be rectified by more information about the problem at hand. However, what of problem ambiguity? What of when there exists a 'state of having many ways of thinking about the same circumstances or phenomena'? (Feldman, 1989, 5). What happens when such ways of thinking are not seen as reconcilable, and thus, may generate vagueness, confusion and stress? Here more accurate information may reduce uncertainty but does not reduce ambiguity. This is because although related, 'ambiguity' differs from 'uncertainty'. Whereas uncertainty refers to an inability to precisely predict something through inexactness or ignorance, ambiguity may be thought of as ambivalence. Consequently, in situations of ambiguity, neither problem identification nor solution specification can be so readily understood as 'applied problem-solving'. Rather, they become entangled in 'the messy realities of the public policy process' (Howlett et al., 2009, 29). Here a conventional

DOI: 10.4324/9781003413608-1

understanding of policy making fails as problem identification is rendered inconclusive and solution formulation is left irresolute. Thus, 'more information is not the answer. The key is to understand how information is presented' (Zahariadis, 2003, 21). In attempting to better understand the policy process we must thereby 'begin with the most important of all limits to high ambition. All our talk of "making" public policy, of "choosing" and "deciding", loses track of the home truth ... that politics and policy making is mostly a matter of persuasion' (Goodin et al., 2006, 5).

Comprehending the 'messy realities' of the policy process thus involves attending to how policy persuasion works. This entails an investigation as to how the persuasive power of representation helps constitute the reality addressed by policy. In the context of policy studies, it may thereby be conjectured that 'representation is not a mirror of reality, but reality is an attribute of representation' (Wagenaar, 2011, 59). By envisaging the policy process from this perspective, '[p]olicy work, then, has to do with making meaning, and, in particular, with managing a variety of meanings' (Colebatch, 2009, 129). Enhancing knowledge of the policy process thus necessitates an understanding of how reality is represented in policy debates. Put simply, it requires attention to how 'meaning making' operates both in and through the policy process.

Interpretive Policy Analysis

This focus on the role of meaning making requires an appreciation that the reality both constituted and addressed by a policy cannot be understood simply through familiarity with facts alone. Instead, the reality of a policy involves a 'perceptual interpretive element' (Kingdon, 1984, 115) wherein 'sense making is an historically and socially contextualized process' (Yanow, 2006, 10). In this 'interpretive approach',

> [t]he meaning of things is neither natural or inevitable, but instead is socio-culturally specific. Meaning is a social product precisely because by acting in certain ways, individuals demonstrate their commitment to classifying a situation along particular lines.
> (Inglis and Thorpe, 2012, 117)

It is through such meaning making processes that representations of reality are constructed, and the 'persuasion' of policy work gets done. An interpretive approach to policy analysis emphasises an understanding of policy realities by 'recognizing that objectivity typically means that we converse with people who agree with our standards of comparison' (Fischer, 2009, 153). It is in this sense that 'rather than asking the question "What are the costs of a policy?" the practitioner of interpretive policy analysis asks instead, "What are the meanings of a policy?"' (Miller et al., 2000, v).

Having emerged during the early 1990s, academic literature focused on this 'interpretive turn' to policy analysis continues a steady path of growth (Wagenaar, 2011; Yanow and Schwartz-Shea, 2015; Boswell, 2016; Rhodes, 2018; Durnová, 2022). Such work seeks to unpack the 'black box' of how the apparent objectivity of the 'reality' addressed by policy may be constituted by representations of that world in policy formulation activity. Thus, '[i]nterpretive policy analysis goes behind the existing beliefs and their communication to examine how they came to be adopted' (Fischer and Gottweis, 2012, 18). This work endeavours to answer complex questions such as '[w]hat are the various ways in which we make sense of public policies? How do policies convey their meanings? Who are the "readers" of policy meanings? To what audiences do policies "speak"?' (Yanow, 1996, ix).

Much research within interpretive policy analysis presumes the policy process as an arena of struggle for dominance and control over issues of contested meaning. This thread of adversarialism is evidenced in the variety of conceptual frameworks dominating the terrain, be they in the form of: 'conflicts over language' in *discourse coalitions* and *narrative analysis*; the 'constant struggles' of *policy paradoxes*; the 'us–them' of *interpretive communities*; the 'ideological struggles' of *critical discourse analysis*; the 'insiders/outsiders' of *poststructural policy analysis*; or the 'intractable controversies' of *policy frames*. Even where such adversarialism is not centre stage, its existence is pre-supposed by foregrounding post-positivist methods of conflict avoidance, most prevalently in the form of normatively-orientated 'collaborative planning' or 'deliberative practice'.

Significantly less work in interpretive policy analysis has focused on understanding how meaning making operates through processes of policy persuasion in the absence of dispute or explicit attempts to avoid conflict. Although both Myerson and Rydin (1996) and Yanow (1996) provide laudable insight into how forms of representation may defuse disagreement in policy debates, their work falls short of supplying a detailed explanation of the mechanisms through which meaning making may constitute the reality that resolves problematic policy ambiguity. Thus, there is a knowledge deficit on how meaning making in the policy process facilitates the resolution of ambiguity and the dissolution of ambivalence through unanimous support for a policy proposal where there exists substantial potential for dispute. Indeed, there is a considerable gap in our understanding of how meaning making activities may suspend rather than resolve, potential disagreement on differences of interest. Likewise, there exists a lacuna in our appreciation of how meaning making may deflect attention from the discussion of logical inconsistencies in policy proposals, rather than confront them. Lacking is a comprehension of how such meaning making operates as a persuasive way to reduce potential friction consequent on the vagueness, confusion and stress engendered by problematic policy ambiguity. What is missing is a clear grasp of how the resolution of problematic policy ambiguity can occur rapidly and without apparent contest so that policy change 'just makes sense'. This book directly addresses such knowledge deficits.

One means to do so may be to wield the analytical tools developed by Foucauldian-inspired studies of *governmentality* (Dean, 2010; Foucault et al., 2010). Such work has demonstrated the difficulty of thinking beyond premises and how the 'common sense' of planning can disguise the contingency upon which such assumptions rest. However, in keeping with Foucault's broader programme, governmentality is mainly concerned with how actors' perceptions are shaped in ways that prescribe how they think they should behave. Much of this work is associated with issues of self-responsibilisation and is directed at critiques of neoliberalism (Raco and Imrie, 2000; Bresnihan, 2019), or the tactical exploitation of strategic objectives to create informal modes of planning that challenge institutions captured by neoliberalism to fashion fluid forms of 'transactive governmentality' that temporarily fracture the grip of market logics (Certomà and Notteboom, 2017). Resonant with such studies is the work of *post-political* thinkers whose research is normally positioned as a critical counterpoint to the tentacular reach of neoliberalism (Swyngedouw, 2009; Metzger, 2018). At first blush it may seem that such post-political work provides a useful theoretical approach to explaining how ambiguity is resolved by suspending potential disagreement over novel policy concepts. However, this vein of research is largely grounded in the view that many mainstream planning approaches are purposefully employed to diffuse opposition towards neoliberalist worldviews and ultimately perpetuate the dominance of capitalist market dynamics and/or the control of elite interest groups (Lennon, 2019). Often with a nod to the work of Chantel Mouffe (2005, 2013), this work usually advocates dissent and resistance. Thus, while nuance may be evidenced in the analysis, employing a *post-political* lens risks predetermining the explanation through the biases inherent to the analytical framework such that the space required for innovative interpretation is constrained.

Accordingly, the analysis presented in what follows works 'up from' the data rather than imposing an established interpretive framework 'onto' the material. As is progressively detailed for the reader as they move through the book, this necessitates drawing on a range of authors rarely consulted in policy and planning academia. These include philosophers, sociologists and cultural theorists such as Paul Ricoeur, Erving Goffman, Kenneth Burke and Roland Barthes. From this, the idea of 'Policy Entitlement' is formulated to explain the interpretation, resonance and modality of new policy concepts that suspend dispute across a range of agendas where significant potential for disagreement exists. Ontologically, this work is premised on a 'constrained idealist' perspective from which 'the existence of an external world places both constraints and opportunities on the reality-constructing activities of social actors, but regard[s] social constructions as having a high level of autonomy from it' (Blaikie, 2010, 17). Harmonising with this is an epistemology of 'social construction-ism' wherein knowledge is neither revealed from an external reality, nor formulated by reason independent of such a reality. Rather, knowledge is conceived as 'the outcome of people having to make sense of their encounters with the physical world and with other people' (Blaikie, 2010, 22).

Reflecting these perspectives, a discourse-centred interpretive approach is employed. Here, meaning is understood as comprising interpretations of signification, significance and applicability. This entails an exploration of the context-dependent constitution of meaning, and consequently, the potential implications of such meaning(s). In so doing, the book engages an investigation of how a policy's reality is discursively constructed. This demands attention to the role played by symbolic language, acts and objects as carriers of meaning in structuring the processes that give rise to ontological suppositions and their associated epistemological assumptions. Through a hermeneutic analysis of meaning in context, the idea of 'Policy Entitlement' that is proposed focuses on how the interpretation of a policy concept through the prism of prevailing rationalities influences perspectives on what it is deemed to signify and its traction in policy debates.

Outline of the Book

The first chapter sets the foundation for the book by profiling the inter-pretive approach adopted. Focus is placed on how intersubjectivity accentuates the production and communication of meanings through varying degrees of collective interpretations regarding signification, sig-nificance and applicability. The theoretical tools assembled and discussed in this chapter centre on discourse analysis. Specific consideration is given to the role played by symbolic language, acts and objects and the phenomenon of 'policy myths'. This chapter thus provides the groundwork for what follows.

The book uses a case study method to inform and illustrate the interpretive approach underpinning the theory developed in the book. Chapter 2 supplies the necessary background to help the reader position the material drawn on in subsequent chapters when progressively building towards a theory of 'Policy Entitlement'. It is in this context that this short chapter describes the rapid emergence, evolution and formal adoption of the 'green infrastructure' approach in Ireland between November 2008 and November 2011.

Chapter 3 builds on the contextual foundation provided in the previous chapter to probe the motivations for the introduction of the term 'green infrastructure' (GI) and the influence this is perceived to have exerted in planning policy formulation. This chapter identifies the process of problematisation and solution specification through forging a 'narrative of necessity'. Specifically discussed is the way GI was interpreted and employed to fashion an identifiable narrative centred on the importance of green spaces to society's physical, social and economic development, as well as to environmental sustainability.

Whereas previous chapters identified a theoretical knowledge deficit and supplied novel empirical material, Chapter 4 offers the first theoretically innovative contribution of the book. This chapter examines and explains how the interpretation of what GI means engendered particular ontological and epistemological perspectives regarding the term's signification, significance and applicability. By focusing on language as a carrier of meaning, this chapter considers how the naming process may be conceived as a practice of reality construction. This is undertaken by deploying theoretical insights from the philosophy of language (Barthes, 2009[1957]; Beardsley, 1981[1958]; Black, 1962; Boyd, 1993; Richards, 1965[1936]), linguistics (Lakoff and Johnson, 2003; Semino, 2008), hermeneutics (Burke, 1966; Gadamer, 2004; Ricoeur, 2002[1975]) and epistemology (Berger and Luckmann, 1966; Schiappa, 2003). Specifically analysed are how the attributes of the term 'green infrastructure' exerted influence on the way it was interpreted. Subsequently considered is how such influence generated certain signifying effects that stimulated views on GI's pertinence and benefit for landuse governance activities. Accordingly, the role played by such meanings in shaping perspectives of green space planning is investigated. Through a concentration on the micro-processes of interpretation, this chapter isolates, examines and explains the processes operative at the first level of

'Policy Entitlement'. As such, it provides the base upon which subsequent chapters develop the theory.

Chapter 5 extends the analysis of the previous chapter by investigating how the naming effects prompted by GI's *entitlement* facilitated its resonance with the rationalities prevalent in the practices of planning and allied disciplines. This is undertaken by critically examining the way such *rationality resonance* both influenced, and was influenced by, assumptions in the arena of planning policy formulation. The supposed 'practice accord' of the GI approach with existing linguistic and procedural characteristics of Irish planning activities is first considered. Subsequently, an exploration of the perceived 'functional' and 'enunciative' advantages of GI is provided. Following this, an inquiry is undertaken on the promulgated 'functional advantage' of GI. By examining the perceived resonance of GI with the prevailing rationality of Irish planning practice, this chapter expands the theoretical interpretation by determining, scrutinising and clarifying the processes operative at the second level of 'Policy Entitlement'.

The discussion of Chapter 6 extends that of previous chapters by moving beyond the GI discourse and investigating the uses to which the narrative was put. The chapter elucidates the processes facilitating the emergence and evolution of a coalition of GI advocates. It then examines how the particularities of GI's interpretation prompted a narrative form that enabled those with varying and often diverging interests to unite in advocating the GI concept. This material is used to help inform, substantiate and illustrate an hypothesis of *narrative modality*. The term 'narrative modality' is used here to describe the proliferation of the GI narrative in both the frequency of its use by a multitude of different agents and the scope of issues it was deemed to address. In studying such narrative modality, this chapter expands the theoretical interpretation by distinguishing, dissecting and scrutinising the processes operative at the third level of 'Policy Entitlement'.

The book's final chapter outlines the theory of 'Policy Entitlement' informed by, but abstracted from, the interpretative policy analysis conducted in the foregoing chapters. This chapter presents an account of how forms of representation and interpretation may appear to resolve issues of problematic policy ambiguity that simultaneously provide a seemingly singular direction for different actors with different agendas. The theoretical contributions of this approach are then profiled. In coming full circle, the chapter concludes by detailing how careful attention to the role of meaning making addresses identified gaps in our understanding of the planning policy process.

References

Barthes, R. 2009[1957]. *Mythologies*. London, England, U.K., Vintage.

Beardsley, M. C. 1981[1958]. *Aesthetics: Problems in the Philosophy of Criticism*. Indianapolis, Indiana, U.S.A., Hacket Publishing Company.

Berger, P. and Luckmann, T. 1966. *The Social Construction of Reality: A Treatise on the Sociology of Knowledge*. London, England, U.K., Penguin Books.

Bevir, M. and Rhodes, R. A. W. 2015. *Routledge Handbook of Interpretive Political Science*. Abingdon, England, U.K., Routledge.

Birkland, T. A. 2015. *Introduction to the Policy Process*. London, England, U. K., M.E. Sharpe.

Black, M. 1962. *Models and Metaphors*. Ithaca, New York, U.S.A., Cornell University Press.

Blaikie, N. 2010. *Approaches to Social Enquiry*. Cambridge, England, U.K., Polity Press.

Boswell, J. 2016. *The Real War on Obesity: Contesting Knowledge and Meaning in a Public Health Crisis*. Basingstoke, England, U.K., Palgrave Macmillan.

Boyd, R. 1993. Metaphor and theory change: what is 'metaphor' a metaphor for? In: Ortony, A. (ed.) *Metaphor and Thought*. Cambridge, England, U.K., Cambridge University Press.

Bresnihan, P. 2019. Revisiting neoliberalism in the oceans: governmentality and the biopolitics of 'improvement'in the Irish and European fisheries. *Environment and Planning A: Economy and Space*, 51, 156–177.

Burke, K. 1966. *Language as Symbolic Action*. Berkley, California, U.S.A., University of California Press.

Certomà, C. and Notteboom, B. 2017. Informal planning in a transactive governmentality. Re-reading planning practices through Ghent's community gardens. *Planning Theory*, 16, 51–73.

Colebatch, H. K. 2009. *Policy*. Berkshire, England, U.K., Open University Press.

Dean, M. 2010. *Governmentality: Power and Rule in Modern Society*. London, England, U.K., Sage.

Durnová, A. 2022. Making interpretive policy analysis critical and societally relevant: emotions, ethnography and language. *Policy & Politics*, 50, 43–58.

Feldman, M. S. 1989. *Order Without Design: Information Production and Policy Making*. Stanford, California, U.S.A., Stanford University Press.

Fischer, F. 2009. *Democracy and Expertise: Reorienting Policy Inquiry*. Oxford, England, U.K., Oxford University Press.

Fischer, F. and Forester, J. (eds.) 1993. *The Argumentative Turn in Policy Analysis and Planning*. London, England, U.K., Duke University Press.

Fischer, F. and Gottweis, H. (eds.) 2012. *The Argumentative Turn Revisited: Public Policy as Communicative Practice*. London, England, U.K., Duke University Press.

Foucault, M., Davidson, A. I. and Burchell, G. 2010. *The Government of Self and Others: Lectures at the Collège de France 1982–1983*. New York City, New York, U.S.A, Palgrave Macmillan.

Gadamer, H.-G. 2004. *Truth and Method*. London, England, U.K., Continuum International Publishing Group.

Goodin, R. E., Rein, M. and Moran, M. 2006. The public and its policies. In: Goodin, R. E., Rein, M. and Moran, M. (eds.) *The Oxford Handbook of Public Policy*. Oxford, England, U.K., Oxford University Press.

Hill, M. and Varone, F. 2021. *The Public Policy Process*. Abingdon, England, U.K., Routledge.

Inglis, D. and Thorpe, C. 2012. *An Invitation to Social Theory*. Cambridge, England, U.K., Polity Press.

Kingdon, J. W. 1984. *Agendas, Alternatives and Public Politics*. New York City, New York, U.S.A., Harper Collins.

Lakoff, G. and Johnson, M. 2003. *Metaphors We Live By*. London, England, U.K., University of Chicago Press.

Lennon, M. 2019. Planning's position in the 'hollowing-out' and 'filling-in' of local government in Ireland. *Planning Theory & Practice*, 20, 612–618.

Metzger, J. 2018. Postpolitics and planning. In: Gunder, M., Madanipour, A. and Watson, V. (eds.) *The Routledge Handbook of Planning Theory*. Abingdon, England, U.K., Routledge.

Miller, M. L., Van Maanen, J. and Manning, P. K. 2000. Series editor's introduction. In: Yanow, D. *Conducting Interpretive Policy Analysis*. London, England, U.K., Sage Publications.

Mouffe, C. 2005. *On the Political*. Abingdon, England, U.K., Routledge.

Mouffe, C. 2013. *Agonistics: Thinking the World Politically*. London, England, U.K., Verso.

Myerson, G. and Rydin, Y. 1996. *The Language of Environment: A New Rhetoric*. London, England, U.K., UCL Press.

Raco, M. and Imrie, R. 2000. Governmentality and rights and responsibilities in urban policy. *Environment and Planning. A*, 32, 2187–2204.

Rhodes, R. A. W. 2018. *Narrative Policy Analysis: Cases in Decentred Policy*. London, England, U.K., Palgrave Macmillan.

Richards, I. A. 1965[1936]. *The Philosophy of Rhetoric*. Oxford, England, U.K., Oxford University Press.

Ricoeur, P. 2002[1975]. *The Rule of Metaphor*. Oxford, England, U.K., Routledge.

Schiappa, E. 2003. *Defining Reality*. Edwardsville, Ilinois, U.S.A., Southern Illinois University Press.

Semino, E. 2008. *Metaphor in Discourse*. Cambridge, England, U.K., Cambridge University Press.

Swyngedouw, E. 2009. The antinomies of the postpolitical city: in search of a democratic politics of environmental production. *International Journal of Urban and Regional Research*, 33, 601–620.

Wagenaar, H. 2007. Interpretation and intention in policy analysis. In: Fischer, F., Miller, G. J. and Sidney, M. S. (eds.) *Handbook of Public Policy Analysis: Theory, Politics and Methods*. Boco Raton, Florida, U.S.A., CRC Press.

Wagenaar, H. 2011. *Meaning in Action: Interpretation and Dialogue in Policy Analysis*. Armonk, New York, U.S.A., M.E. Sharpe.

Yanow, D. 1996. *How Does a Policy Mean? Interpreting Policy and Organisational Actions.* Washington DC, U.S.A., Georgetown University Press.

Yanow, D. 2000. *Conducting Interpretive Policy Analysis.* London, England, U.K., Sage Publications.

Yanow, D. 2006. Thinking interpretively: philosophical presuppositions and the human sciences. In: Yanow, D. and Schwartz-Shea, P. (eds.) *Interpretation and Method: Empirical Research Methods and the Interpretive Turn.* London, England, U.K., M.E. Sharpe Inc.

Yanow, D. and Schwartz-Shea, P. 2015. *Interpretation and Method: Empirical Research Methods and the Interpretive Turn.* Abingdon, England, U.K., M. E. Sharpe.

Zahariadis, N. 2003. *Ambiguity and Choice in Public Policy: Political Decision Making in Modern Democracies.* Washington DC, U.S.A., Georgetown Press.

1 Interpreting Policy Discourses

Studying the Policy Process

Public policy scholarship has 'a long history and a short past' (Howlett et al., 2009, 17) wherein there is a long tradition of investigating the actions of government (Moran et al., 2009; Ansell and Torfing, 2016), while the systematic examination of policy using academic frameworks dates back just seven decades (Hill and Varone, 2021). However, over this comparatively short period, interest in the study of the policy process has grown significantly and is characterised by a considerable number of overlapping, yet discrete, perspectives (Weible and Sabatier, 2018; Cairney, 2019). Thus, public policy research may be understood as 'a loosely organized body of precepts and positions rather than a tightly integrated body of systematic knowledge, more art and craft than genuine "science"' (Goodin et al., 2006, 5).

This has led some to bemoan the field of policy studies as 'a babel of tongues in which participants talk past rather than to one another' (Bobrow and Dryzek, 1987, 4). Navigating this diversity may be rendered less onerous through an appreciation that public policy research is largely partitioned between knowledge *for* policy and knowledge *of* the policy process (DeLeon and Martell, 2006; Hogan and Howlett, 2015). Knowledge 'for' policy principally refers to knowledge produced through empirical evaluation and normative assessment, thereby concerning the ex-ante marshalling of information to assist policy makers and the ex-post appraisal of initiative implementation (Colebatch, 2009; Hill, 2009). In contrast, knowledge 'of' the policy process is centred on the 'why' and 'how' of policy making (Nowlin, 2011; Weible and Workman, 2022). It is less focused on normative appraisal and more concerned with 'finding out why governments pay attention to some problems and not others (agenda setting), why policy changes or remains stable across time, and where policy comes from' (Smith

DOI: 10.4324/9781003413608-2

and Larimer, 2009, 6). Given that the focus of this book is on examining the policy process as both constituting and revealing rationalities through the varying interpretations of a policy's meaning, what follows is primarily focused on the 'why' and 'how' of policy making. This book is thereby concerned with knowledge 'of' the policy process.

A diversity of theoretical approaches has been applied to studies 'of' the policy process. These include, but are not limited to, conceptual frameworks borrowed from: historical institutionalism (Rayner, 2009; Sanders, 2006; Thalen, 1999; Isaksen and Jakobsen, 2019); sociological institutionalism (Hay, 2006; Scott, 2008; Béland, 2019); network focused analysis (Compston, 2009; Kenis and Schneider, 1991; Murdoch, 2000; Rhodes, 1997, 2006); work informed by science and technology studies (Burgess et al., 2000; Collins et al., 2009; Latour, 2005; Rydin and Tate, 2016); Marxist inspired approaches (Howlett, 2010; O'Sullivan, 2003; Waldron, 2019); work guided by varying forms of structuration theory (Crowley, 2006; Delmas, 2002; O'Dwyer and O'Sullivan, 2015); and post-structuralist analysis (Gottweis, 2003; Griggs and Howarth, 2019). Different forms of interpretive analysis have been employed within these conceptual frameworks to supply nuance in the deciphering and describing of what is done by whom and why. Central to these efforts has been the use of discourse analysis.

Discourse Analysis

Although a broad church of many different perspectives, discourse theorists are united by a desire to describe, understand and explain particular phenomena in the context of their occurrence rather than establish generalisations or test universally applicable hypotheses. They maintain that it is not reality in an observable or testable sense that shapes social consciousness and action, but rather it is the ideas, beliefs and values that discourses evoke about the causes of satisfactions and discontents that mould comprehension and intent. Thus, in contrast to empiricist epistemologies, discourse theorists are preoccupied with exploring how, in what context and for what reasons discourses are constructed, contested and changed by whom and when. As an implication of this approach, discourse analysis theories start from the assumption that all forms of human communication, be it conveyed via language, objects, acts or practices, are socially meaningful and that these meanings are shaped by social, cultural and political conditions of period-specific contexts.

Fundamentally, discourse theories hold that all knowledge is discursively constructed through shared understandings of context-specific meaning. In this manner, a discourse can be appreciated as a 'shared way of apprehending the world' (Dryzek, 2005, 9). Discourse theorists do not contend that there is no world external to discourse, but instead argue that comprehensions of this world are mediated by discourse. It is this mediating process that prompts the perceptions of objectivity that are conceived to constitute what counts as real. Thus,

> [d]iscourse theory does not dispute in any way the realist assertion that matter exists independently of our consciousness, thoughts and language. The contention is that nothing follows from the bare existence of matter. Matter does not carry the means of its own representation ... Rather, intelligible social forms are constructed in and through different discourses. Hence, a particular piece of land can be constructed as habitat for an endangered species by a group of biologists, a recreational facility by the urban population, fertile farm land by local farmers, or a business opportunity by urban developers.
>
> (Torfing, 2005, 18)

Therefore, discourse analysis refers to the process of scrutinising the practices employed in the construction of discourses and the influences of discursively mediated interpretations. It follows that those engaged in discourse analysis examine the broad spectrum of linguistic and non-linguistic material that enables interpreting subjects to experience the world. Through reference to this approach, 'discourse' is here understood to be more than the 'mode of talking' synonymous with common parlance. Rather, it is conceived as a specific and cohesive ensemble of ideas, concepts and categorisations that are produced, reproduced and transformed in a particular set of practices and against the background of a specific social, temporal and spatial context (Epstein, 2008; Hajer, 1995). Applied to the formulation of policy concerning the interface between society and the environment, Feindt and Oels note that,

> [t]aking a discursive perspective allows one to understand how 'nature' and 'the environment' are continuously 'produced' through environmental policy making, planning, research and development as well as through everyday practices. It also allows one to ask if the environmental policy is about nature and the environment at all or rather about the redistribution and reconfiguration of power in the name of the 'environment'.
>
> (Feindt and Oels, 2005, 163)

With a particular focus on environmental planning policy, this suggests that examining 'how' agents use discourse in providing clarity of meaning to landuse policy ambiguity furnishes a way to investigate 'how' the context contingent interpretations of such clarification influences the principles of landuse governance. Additionally, attention to 'what' is communicated through discourse permits an examination of how such rationalities evolve over time. This is primarily achieved through an appreciation of the ways in which discourses function to regularise how a particular issue is perceived both ontologically and epistemologically, and thus how the basic principles of social action are structured in relation to it. Hence, discourses have formative power in configuring shared understandings and human interactions with both the social and physical worlds. As such, realities are never understood simply through familiarity with facts alone. Rather, realities are conceived to involve a 'perceptual interpretive element' (Kingdon, 1984, 115) which is organised by particular discourses that transmit context specific meanings that both constitute, and are constituted by, systems of knowledge. Discourse theorists assert that questions of truth and falsity are therefore not resolved by a theory-independent world of phenomena. Instead, such questions are seen as relative to the standards of authentication established by particular systems of knowledge which are related to specific places during certain periods (Foucault, 1972[1969]). Discourse analysis thereby shifts the focus from objective truths to a 'will to truth' (Foucault, 1976, 1977; McNay, 1994; Sheridan, 1980). This reflects the complex set of relationships between knowledge that is produced during a particular period and the rules by which new knowledge is generated (Hacking, 2002). Accordingly, within a particular period, discursively associated meanings construct similarities in the systems of knowledge operative at a conceptual level, despite often dealing with different subject matters (Dreyfus and Rabinow, 1982; Smart, 1985).

Used parsimoniously as a backdrop to an analysis of specific discourses, rather than as the object of such an analysis, this view on how standards of authentication are context dependent can provide a means for understanding how some concepts gain traction in debates among parties schooled in specific modes of thought (Ameel, 2020). This can be seen to reinforce Schmidt's (2011) contention that the ideas in a discourse must not only 'make sense' within a particular meaning context, but that the discourse itself must be patterned according to a given 'logic of communication,' following rules and expressing ideas that are socially constructed and transmitted within a given discursive setting. Thus, employing discourse analysis facilitates an examination of how meaning making operates to produce policy realities resonant with the prevailing presuppositions of knowledge legitimacy in a particular context. In so doing, it enables an investigation into how meaning making may provide clarity in moments of problematic policy ambiguity. However, examining how discourse may be employed in this way implies an effort to denaturalise what is assumed as 'the truth about a reality'. The task is thus not to evaluate whether statements are true or false, but rather to investigate 'how' such 'truths' are mobilised. In other words, the job is to study 'how' meaning making engenders the apparent legitimacy of a particular approach to policy making. For as noted by Epstein,

> [t]he 'truth' is potent. Its power is wielded in particular discursive economies of power. Thus, it becomes necessary to assert the relativity of truth claims and to consider them in relation to the particular configuration of power relations within which they obtain. More generally, studying discourses is a means to taking a critical step out of what the discourses actually say in order to observe what they do.
>
> (Epstein, 2008, 13)

Consequent on these 'discursive economies of power', actors occupied with discursive activity are positioned relative to the subject of that activity. Discourses thereby part-constitute the identities of social actors by creating particular 'subject positions'. Put simply, discourses specify the power and positions from which social actors can communicate and act with influence. The power relations inherent in the use of discourses may be both constraining and enabling to the actors who engage in their use. Foucault (1972[1969]) elucidates this idea by arguing that who says what, where, when and how, and with what influence, is shaped through the evolution of discursive rules that constitute 'enunciative modalities'. He therefore places emphasis on the need to investigate the many ways in which different actors are bestowed the mandate to speak authoritatively on issues consequent on their positions.

One limitation on the ability to authoritatively pronounce on an issue is the capacity to present arguments grounded in what are perceived as valid forms of knowledge (Steffek, 2003, 2009; Birkland, 2015). As legitimate governance in modern western democracies is set against the backdrop of an historical legacy wherein justifiable action is seen to follow sequentially from 'objective' knowledge acquisition (Fry and Raadschelders, 2008; Lennon and Tubridy, 2022), the possession of valid (objective) knowledge is a key determinant on the ability to authoritatively pronounce on an issue. Such valid forms of knowledge habitually partition the world into apparently self-evident dichotomies of true and false; objective and subjective. This 'naïve realism' is the ground of modernist rationalities that view the universe as comprising 'autonomous actors and an independent reality' (Wagenaar and Cook, 2003, 140). In this way, 'the general state of reason' (Foucault, 1972[1969]) delimiting the legitimacy of knowledge in modern western democracies, and thus the power to govern in such contexts, is set in an ability to underpin governing activity by an appeal to knowledge which appears to have been conceived in accordance with the rules of such modernist rationalities (Aronowitz, 1988; Gane, 2004; Weber, 1922). Flyvbjerg (1998) extends this idea by showing that it is the 'appearance' of such rationalities rather than a genuine concern with their use that is important in power-imbued governing activity.

This 'knowledge dependence' (Gottweis, 2003, 256) of governing activity has important implications for the currency of a new planning policy 'approach', where an approach is understood as a broadly-shared perspective on how a particular set of problems should be addressed. Specifically, as the perceived legitimacy of landuse policy generally relies on reference to modernist rationalities (Flyvbjerg, 1998; Rydin, 2003; Phelps, 2021), the capacity of a proposed landuse policy to resonate with prevailing interpretations as to what comprises valid knowledge is likely to exert significant influence on its adoption by those positioned within planning and allied professional disciplines. Furthermore, those in a position to enunciate such knowledge are likely to assume identities constituted by power relationships, and enjoy relative to others the ability to identify, control and legitimise the very issues taken to be the subjects of deliberation. In this sense, 'the question of who should have the authority to make definitional decisions amounts literally to who has the power to delineate what counts as Real' (Schiappa, 2003, 178).

Those working in the field of interpretive policy analysis have directed attention to the importance of 'causal stories' (Stone, 1989, 1997) in this process by furnishing the 'collective centring' (Hajer and Laws, 2006, 260) that allows constellations of actors to coalesce around a series of associated storylines. Such 'collective centring' of different interests has been widely described as a 'discourse coalition' (Hajer, 1993; Fischer, 2003; Epstein, 2008; Runhaar, 2009; Wagenaar, 2011). These coalitions comprise the well of support for a policy. Therefore, the size of a discourse coalition and 'who' it includes is likely to affect traction of a new policy approach in decision-making activities. Moreover, the composition of such a coalition may significantly influence the way a policy evolves, the pace with which it ascends the decision agenda, and both the degree and speed with which it is subsequently adopted.

Discourse Coalitions

Based upon research concerning acid rain debates in Great Britain and the Netherlands during the 1980s, Hajer theorises that 'discourse coalitions',

> ... are defined as the ensemble of (1) a set of story-lines; (2) the actors who utter these story-lines; and (3) the practices in which this discursive activity is based. Story-lines are here seen as the discursive cement that keeps a discourse-coalition together. The reproduction of a discursive order is then found in the routinization of the cognitive commitments that are implicit in these story-lines.
>
> (Hajer, 1995, 65)

As the keystone in this discourse coalition hypothesis, storylines have a number of essential qualities. Hajer (1995, 63) outlines these as follows:

> (I) They function in distilling the complexity of a problem and often involve discursive closure
> (II) They create easily comprehensible solution possibilities that also frequently involves discursive closure
> (III) They have a ritualistic dimension which through repetition gives permanence and perceived validity to their content
> (IV) They allow different actors to expand their appreciation and discursive proficiency on an issue beyond their disciplinary expertise or experience

In this theory, storylines are conceived as forming tropes or shortcuts into broader narrative schemes that configure events and actions into a unified order which identifies the larger patterns to which they contribute. This organising process operates by connecting diverse phenomenon and stipulating the causal chain of effects that each phenomena has on each other (Kaplan, 1993; Roe, 1994; Dillon and Craig, 2021). Therefore, storylines not only convey meaning, but also offer those who subscribe to them a way of perceiving the phenomena under examination (Fischer, 2003; Lejano et al., 2013). In other words, storylines orientate interpretations, and in doing so, they help constitute reality for those who subscribe to them (Paltridge, 2006; Finch, 2021).

Of particular note is Hajer's contention that the power of storylines to form discourse coalitions derives from their capacity to facilitate 'discursive affinities' (Hajer, 1993, 1995, 2005). These are envisaged as separate elements that have similar cognitive or discursive structures and so tacitly suggest a logical mutuality. Such affinities do not primarily refer to agents and their intentions, but instead allude to the influence of discursive formats on the perception of reality. For example, an agent may not comprehend the technical details of an argument, but may be confident in asserting that it 'sounds right' (Hajer, 1995, 67). Furthermore, Hajer theorises that in the case of a particularly strong affinity, discursive elements not only resemble one another, but an exchange of terms or concepts may also exist. He terms this phenomenon 'discursive contamination' (Hajer, 1995). Consequently, discursive affinities and contaminations may be thought to function in clustering interpretations of meaning that share a broadly-aligned logic rather than an issue-specific assertion. In this way, the various agents comprising a discourse coalition can be conceived as capable of forming associations in their support for the reasoning upon which an array of discursive affinities and contaminations are able to successfully operate. This is achieved by permitting latitude in interpretation of the particular problems or policies perceived to be addressed by the expressions that prompt discursive affinities and contaminations. Discursive affinities and contamination may thereby be seen as both reflecting and constituting the reality on which the rationalities of policies are based. In the case of landuse planning, wherein the perceived legitimacy of policies is generally reliant on reference to modernist rationalities (Throgmorton, 1993; Hawkesworth, 2012; Heazle and Kane, 2015), the influence of discursive affinities and contaminations in sustaining and expanding a discourse coalition among planners and allied professionals is likely to be predicated on their ability to resonate with 'the traditional view' (In't Veld, 2009, 121) of a 'technical-rational model' (Owens et al., 2004, 1945) of knowledge production conceived as operative within planning practice (Flyvbjerg, 1998; Rydin, 2003, 2007; Couch, 2017; Hall, 2019). Indeed, Richardson (1996, 282) notes how '[p]olicy analysts and planners have frequently claimed that their work is based on rationality and objective reason. "Facts" supporting arguments in policy making are generally supported by such claims to rationality.' Studying 'what', 'why' and 'how' certain discursive affinities and contaminations connect different policies may thus be employed to identify the rationalities underpinning the principles of landuse governance embodied in a new planning policy approach.

Whereas Hajer's discourse coalitions hypothesis focuses largely on the role of language, a discourse perspective may be 'concerned with any type of signifying practice, that is, any practice that functions as a site for the production of meaning' (Epstein, 2008, 186). It is therefore also important to remain attentive to the role of acts and objects as carriers of meaning in consolidating the policy realities prompted by discursive affinities and contaminations. As such, language, acts and objects can be viewed as 'symbols' that weave a web of signification in structuring the reality both constituted by, and addressed in, policy work. Each such 'symbol is a social convention' (Yanow, 2000, 14) whose meaning is broadly agreed upon but not delineated (Gold and Revill, 2004). Symbols thereby often communicate through connotation rather than denotation (Chandler, 2007; Hébert, 2019). Where these symbols are perceived to connote knowledge legitimated in accordance with accepted disciplinary standards, such as those in landuse planning, they may be conceived as representing factual statements and thereby meet approval (Swaffield, 1998; Ockwell and Rydin, 2006; Lennon and Waldron, 2019). Seen in this light, symbols can offer the medium through which diverse motivations, expectations and values are synchronised to enable accord between numerous interests across a spectrum of issues and institutional contexts. Accordingly, symbolic language, acts and objects may furnish the connotations that 'will to truth' (Foucault, 1976, 55) the facts that enable the formation, maintenance and expansion of a discourse coalition.

Symbolic Language

By attending to the symbolic role of language as both constituting and carrying the meanings engendered in policy work, 'language becomes part of data analysis for inquiry, rather than simply a tool for speaking about an extra linguistic reality' (Shapiro, 1981, 14). Appreciating this constitutive role requires attention to 'what happens when people draw on the knowledge they have about language ... to do things in the world' (Johnstone, 2018, 3). Drawing on such knowledge entails mediating communication through the context-contingent linguistic conventions that supply the pre-conditions for the process of discourse formation (Lemke, 1998, 91). Central to this is the part language plays in the categorisation of experience, and as such, fostering 'mental constructs in a world that has only continua' (Stone, 1997, 378). Yanow (2002) suggests that these constructs are central to the policy process through their influence in structuring perceptions of the reality upon which policy is directed. Categories are important in this respect as they function to structure the interpretation of the content they signify through connotation (Kingdon, 1984; Rhodes, 2018). However, categories are not 'fixed', 'innate' or 'given' phenomena. Rather, classification can be understood to involve an interpretive choice based on conclusions regarding the relative importance of some features over others. Categories emphasise elements deemed commensurate within their delineations and the possible associations between groupings (Johnstone, 2018). As a corollary, categories help silence those elements which they do not deem to be significant (Bowker and Leigh-Star, 1999). In this way, 'it is through categorization that the specific sense of something is constituted' (Potter, 1996, 177). Categories thereby imply certain attributes about that which is classified. Accordingly,

> [l]anguage is capable not only of constructing symbols that are highly abstracted from everyday experience, but also of 'bringing back' these symbols and appresenting them as objectively real elements in everyday life. In this manner, symbolism and symbolic language become essential constituents of the reality of everyday life and of the common-sense apprehension of this reality.
>
> (Berger and Luckmann, 1966, 55)

Given their potential to configure the world in an apparently logical format, language-induced categories offer an important symbolic apparatus open to use by those seeking to locate meaning in situations of problematic policy ambiguity. Be they the product of unintentional evolution or deliberate application, they may function as essential elements in constituting the storylines from which problems and/or policies emerge, as well as how problems and policies are coupled. This is facilitated by their capacity to be drawn on as quotations, references and heuristic devices to partition the world and shape realities. As such, the configuration of categories through language profoundly shapes our view, and in this way, may be used to delineate what can be considered as legitimate knowledge.

However, engendering forms of reality by categories need not be done explicitly. Instead, '[t]he fundamental legitimating "explanations" are … built into vocabulary' (Berger and Luckmann, 1966, 112). Hence, 'the mere act of naming an object or situation decrees that it is to be singled out as such-and-such rather than as something other' (Burke, 1973, 4). The naming process may thereby be conceived as a process of reality construction (Potter, 1996, 82). It is in this context that Burke (1966) advances a 'theory of entitlement' wherein he proposes a reversal of the intuitive understanding that 'words are the signs of things' by suggesting that 'things are the signs of words' (Burke, 1966, 360–361). As explained by Schiappa,

> [t]o 'entitle' something – 'X' – is not only to give X a title in the simple sense of assigning X a name or label, but it is also to give X a particular status. For example, to describe X as 'an object' is to assign X an ontological status somewhat different than labelling X 'an event' or 'a vague feeling'.
>
> (Schiappa, 2003, 114)

Burke's proposition is that naming may 'entitle' reality. It is through this process of entitlement that presuppositions of how something can be known may be stimulated. In cases where an agreed definition of that which is entitled is absent, a term may be employed so variously that it becomes 'underdetermined in meaning and overdetermined in figuration' (Allen, 2000, 2), such that the boundaries between literal and figurative expression are blurred. Here, the frontiers separating exact speech and analogy may become porous as understandings of that which is named tack back-and-forth between connotations and denotation (Barthes, 2009[1957]). In this sense, language may become 'at once literal and figurative, and hence intrinsically metaphorical' (Orr, 2003, 162). Here, ontological and epistemological presumptions may be transferred from familiar concepts onto new abstract ideas whose definition is still in flux (Moran, 1995). New meaning may thereby be acquired by drawing upon existing knowledge of something known and familiar. Schön (1993) has demonstrated how in a policy context this may be observed in the use of metaphors to orient attention towards novel ideas. Metaphors facilitate this as they are both fundamentally conceptual in nature but also grounded in everyday experience (Kövecses, 2002; Knowles and Moon, 2006; Lyon, 2000; Semino and Demjén, 2016). Accordingly, 'the essence of metaphor is understanding and experiencing one kind of thing in terms of another' (Lakoff and Johnson, 1980, 5). Metaphors are thus heuristic devices that comprise the juxtaposition of two superficially dissimilar elements in a single context (Hausman, 2006; Richards, 1965[1936]).

As has been demonstrated by Myerson and Rydin (1996) with respect to environmental policy, although metaphors may initially appear as merely descriptive, they function by directing perception (Black, 1962; Ricoeur, 2002[1975]). While they may offer new insights into phenomena, they may simultaneously help conceal elements of those phenomena (Goatly, 1997; Ortony, 1993; Semino, 2008; Musolff, 2016). Like categories, metaphors emphasise certain aspects of things and obscure others, thereby organising perceptions of reality and suggesting appropriate actions in response to such perceptions. It is their conventionality, tacit knowledge potential and the similarities in their broadly-shared sense of meaning among a community of interpreters that masks the power of metaphors to shape action (Boyd, 1993; Yanow, 1996). Put simply, the power of metaphors resides in their ability to mould action in response to the perceptions of the meanings they provoke (Hart, 2008). In this way, metaphor may be employed to reduce ambiguity through the transference of ontological and epistemological connotations from the familiar onto a new idea. Furthermore, in time, and through frequent use, the connotations of such metaphorical reasoning may evolve into what are increasingly perceived as denoted 'facts' (Barthes, 2009[1957]; Beardsley, 1981[1958]). Here, metaphor may be conceived as providing stability of meaning to problematic policy ambiguity such that it becomes more description than analogy in a 'will to truth' (Foucault, 1976, 55). Thus, metaphors may evolve from models 'of' a situation to models 'for' it. Should the language of the metaphor(s) used facilitate 'discursive affinities' and/or 'contaminations', metaphor may provide a powerful means by which to create, sustain and expand a coalition of support for a policy or series of polices orientated to a reality entitled by language (Schiappa, 2003, 115). Consequently, it is conceivable to think that the strategic use of metaphor may assist the emergence and evolution of a new policy approach.

Symbolic Acts

'Acts stand in a representational relationship to the meanings understood or intended to underlie them' (Yanow, 2000, 74). Thus, acts offer windows into the rationalities which underpin a policy approach. Frequently prominent among these is the act of counting (Evans, 2007; Flyvbjerg, 1998; Hajer, 1995; Sanderson, 2002; Bogenschneider and Corbett, 2021). As noted by Kingdon (1984, 98), quantified information 'acquires a power of its own that is unmatched by issues that are less countable'. Thus, statistics may be employed in a 'will to truth' by asserting a 'claim to "factual" status and the elevation of information into objective knowledge' (Myerson and Rydin, 1996, 21). It is in this way that the act of counting may be conceived as symbolically directed to establishing the 'numinous legitimacy' (Clark and Majone, 1985, 16) of objective scientific inquiry that conveys meaning seemingly independent of those who engage in the measurement exercise. However, the very act of counting can also serve a normative function by implying a need to do something. Hence, in quoting efforts to quantify unemployment in the U.S.A. during the 1930s, Stone (1997, 167) notes how the deed of measuring usually implies 'a need for action, because we do not measure things except when we want to change our behaviour in response to them'. Furthermore, Fischer (2003, 171) suggests that '[b]y establishing recognisable boundaries, counting can normatively function like metaphors'. In this way, counting may be conceived as a symbolic act that helps constitute the reality of a policy approach.

Additionally, quantifying endeavours may be ambiguous, particularly when they concurrently suggest explicit and implicit stories that blur the boundaries separating value and instrumental rationality (Flyvbjerg, 1998). The act of measuring, the methodologies used and the results of such exercises can thereby be patterned in ways that establish the normative validation of a proposal through their integration into storylines of rectitude and rationality (Hannigan, 2014; Clemons and McBeth, 2020). Statistics can thus be employed to enhance the potency of a utopian or dystopian storyline, thereby helping to tacitly or overtly communicate meaning in a way that orientates interpretation and prompts action (Meadows et al., 1972; Dryzek, 2013). Moreover, it may be possible that the ostensible numeracy of a storyline can house or reference an implicit narrative that functions independently of the would-be meaning of the numbers (Jones et al., 2014). For example, Fischer (2003, 172) suggests that 'a hidden message is often transmitted in the very act of counting'. In this scenario, the process of quantification itself may serve as a tacit message signifying that something occurs frequently enough, or is of a sufficient importance, to merit numerical examination, and thus should be taken seriously. In this sense, measuring can be understood as a symbolic act that may be used to shape interpretations of reality and/or provide clarity of meaning in situations of problematic policy ambiguity.

Another symbolic act germane to the emergence of a new policy approach is the act of comparison. Kingdon (1984) identifies how comparison with both the circumstances of another and/or one's past performance may provoke the reconceptualisation of an existing condition as a problem. In this way, comparison may be considered a symbolic act through its role in constituting the meanings that may be employed to direct perception in advancing a policy proposal. Also, just as there is a concern with connoting legitimacy through the scientific objectivity of counting, so too does the symbolic act of comparison suggest a concern for neutrality. Such concern may be particularly strong in the bureaucracy of landuse governance whose legitimacy is reliant on its perceived ability to operate independently of personal or vested interests (Owens et al., 2004; Rydin, 2003; Swain and Tait, 2007).

This attention to an appearance of impartiality in structuring and communicating the validity of knowledge claims has been termed 'stake inoculation' (Potter, 1996, 125). Central to this is the relationship between the identity of those referencing a knowledge claim, those identified as producing such a claim, and that upon which the claim is made. The stake inoculating potentials and properties of such relationships were explored by Erving Goffman (1979, 1981) and elucidated in his theory of 'footing'. Goffman's approach is to refine presumptions on the simple distinction between addresser and addressee by theorising the various roles transcending this dichotomy through proposing a threefold typology of reference. Focused on 'the production or reception of an utterance' (Goffman, 1981, 128), he theorises three discrete roles available in all forms of reference, be they explicitly or implicitly delivered. These are namely the *principal*, whose position the piece of speech is supposed to represent; the *author*, who does the scripting; and the *animator*, who says the words. As such, the theory of footing contends that 'the notions of animator, author, and principal, taken together, can be said to tell us about the "production format" of an utterance' (Goffman, 1981, 145).

These distinctions between *principal, author* and *animator* may be employed in structuring the symbolic act of comparison by exerting influence on the appearance of neutrality through positioning the 'animator' as 'just passing something on' (Potter, 1996, 143). Indeed, 'it is through the paraphernalia of footing that speakers managed their personal or institutional accountability' (Potter, 1996, 122). Several studies have explored the role of comparison in directing the interpretation of problems and policies (Epstein, 2008; Kingdon, 1984; Zahariadis, 2003; Halpern et al., 2021). Nevertheless, instances where the theory of 'footing' has been applied in the analysis of 'how' this process was undertaken are rare in policy and planning research, with such use largely confined to media studies (Goodwin, 2006; Tolson, 2006). However, employed in the examination of how perceptions of impartiality are shaped through the process of comparison, the theory of footing can help elucidate 'how' meaning making activity operates to resolve problematic policy ambiguity.

Symbolic Objects

In addition to the symbolic use of language and acts, the work of remedying problematic policy ambiguity may be assisted through the meanings constituted and carried by symbolic objects. In the context of landuse planning whose focus is inherently spatial, one of the most prevalent objects employed is the map. The symbolic quality of cartography and map use rests on connotations of 'veracity' and 'integrity' (MacEachren 1995, 337). These are specified as the implications of temporal and attributive precision commonly associated with impressions of accuracy in mapping, and the presumption of impartiality in the activities of both scientifically-schooled cartographers (Dorling and Fairbairn, 1997; Wilson, 2017), and more recently those with expertise in geospatial software (Ferrari and Rae, 2019). As such, the plans produced with and through maps facilitate the 'stake inoculation' necessary for the enunciative modalities that favourably position agents within discourses of apparent scientific objectivity. Consequent on such epistemological assumptions, maps may 'connote a directness in representation which prompts users to overlook the fact that maps are representations' (MacEachren, 1995, 339).

Map making is thereby a form of meaning making (Cosgrove, 1999; Daniels et al., 2011) wherein the 'medium of communication is ultimately connected with the message it communicates' (Yanow, 2000, 17). Rather than neutral, maps as symbolic objects and the carriers of meaning possess their own affordances and constraints (Crampton, 2003), which are 'already charged with cultural signification' (Eco, 1976, 267). For example, '[i]n "plain" scientific maps, science itself becomes the metaphor' (Harley, 1992, 241). Nevertheless, such condensation of scientific legitimacy in cartography may not only entail resonance with presumptions of proper planning methods; it may also concurrently involve shaping perceptions of that which is presented (Wood, 1992). This is because '[m]apping is epistemological but also deeply ontological – it is both a way of thinking about the world, offering a framework for knowledge, and a set of assertions about the world itself' (Kitchin et al., 2009, 1).

Thus, maps as symbolic objects embody presumptions on legitimate forms of knowledge and orientate interpretations of the reality they claim to represent. The 'will to truth' of information presented in cartographic form may thereby enable map authors to legitimately expound 'an' interpretation of something as 'the' interpretation via reference to an apparent objective reality grounded in the 'numinous legitimacy' of science (Clark and Majone, 1985, 16). Through the production of symbolic objects, cartography can thus allow map authors to legitimately proclaim the 'facts' of a situation from an advantageous enunciative position via appeal to the seeming objectivity engendered by stake inoculation. Put simply, maps legitimise that which is enunciated.

For a map to connote a 'truth' relative to the suppositions of its audience, the activity of map making must be selective in content (Monmonier, 1991). Selectivity requirements thereby permit the use of maps as devices that channel interpretation by highlighting and discounting the aspects of the reality its authors seek to construct (Blacksell, 2006; Corner, 1999; Evans, 2007; Wilson, 2017). Consequently, maps function similarly to metaphors as both instruments of communication and a means of persuasion (Pickles, 2004; Dodge et al., 2011). As such, the symbolic qualities of maps may be employed to offer clarity of meaning and provide direction for action on issues of problematic policy ambiguity. Key to ensuring that such meaning enjoys traction in policy decision-making is how it enables agents with different agendas to agree on a solution. Important in this process is the role of 'policy myths'.

Myth

In functioning as the carriers of meaning, symbolic language, acts and objects are central to the production of policy myths. These forms of narrative are social constructions embedded in a particular time and place. They offer an account of reality which through allusion to the symbolic qualities of their composite storylines supply 'figures of resolution' (Myerson and Rydin, 1996, 181) to issues of problematic policy ambiguity. This is achieved by presenting clarity of meaning on the identity and nature of problems, as well as suggesting how such problems may be remedied. The term 'myth' is employed here to designate a 'narrative created and believed by a group of people which diverts attention from a puzzling part of their reality' (Yanow, 1996, 191). The idea of myth forwarded in this context is not conceived as an evaluation of a narrative's veracity, as myths are neither true nor false in the empiricist sense. Rather, discernment of their 'truthfulness' is dependent on subscription to their narrative (Bottici, 2007). As such, 'myth' as used here refers to a particular narrative format that facilitates subscription by a broad range of issue-specific interests through proffering apparent commensurability in situations where plausible discrepancies may coincide. Myths achieve this by suspending conflict in 'masking the tensions between or among incommensurable values' (Yanow, 2000, 80) and deflecting attention away from prospective logical inconsistencies or potential incompatibilities in that which is articulated (Charteris-Black, 2009). Consequently, a policy myth may be central to facilitating the emergence, maintenance and expansion of a discourse coalition. This potential may be enhanced by the capacity of myths to implicitly legitimate the actions which their narrative begets. As theorised by Roland Barthes,

> [m]yth does not deny things, on the contrary, its function is to talk about them; simply, it purifies them, it makes them innocent, it gives them a natural and eternal justification, it gives them clarity which is not that of an explanation but that of a statement of fact ... it organizes a world which is without contradictions because it is without depth, a world wide open and wallowing in the evident, it establishes a blissful clarity: things appear to mean something by themselves.
>
> (Barthes, 2009[1957], 169–170)

Myths evolve from the interpretation of meanings communicated via the symbolic language, acts and objects employed in forwarding particular interpretations of problems and policies. By enabling the suspension of potential inconsistencies, contradictions and conflicts, myths thereby offer a means to resolve problematic policy ambiguity.

Summary and Conclusions

As noted by Howlett et al., (2009, 9), 'public policy-making is rarely as simple a matter as either analysts or policy-makers might wish for'. This has led some to exclaim that 'there is no general theoretical framework tying together the study of public policy' (Smith and Larimer, 2009, 15), but rather that such study comprises 'a babel of tongues in which participants talk past rather than to one another' (Bobrow and Dryzek, 1987, 4). Successfully negotiating this babel involves more than assuming that agents are simply 'muddling through' the policy formulation process (Lindblom and Woodhouse, 1992). Rather, it entails an appreciation that '[p]olicies are intentions, the product of creative human imagination' (Goodin et al., 2006, 19). Thus, '[g]iven the staggering complexity of the policy process, the analyst must find a way of simplifying the situation in order to have any chance of understanding it' (Sabatier, 2007, 4). However, the student of policy must be vigilant in seeking to achieve this as elucidating upon the ascension of new policy requires explanation that is 'parsimonious to be sure, but not over simplified' (Greenberg et al., 1977, 1543).

Locating a means to structure such an explanation requires attention to the presumptions on which much policy process work is based. A great deal of this work seeks to derive 'generalizable knowledge and principles that can be applied to achieve policy goals across domains and settings' (Hajer and Laws, 2006, 251). To facilitate the production of such 'generalizable knowledge', much thinking on the policy process presupposes a linear logic wherein new policy is assumed to emerge in a temporal progression from problem identification to solution specification (Wu et al., 2017; Hill and Varone, 2021; Knill and Tosun, 2020; Peters, 2021). However, this position 'does not suggest a way of understanding how policy makers deal with ambiguities and how ambiguity might relate to policy changes' (Hajer and Laws, 2006, 256). Additionally, most policy process theory involves presumptions on adversarial forms of policy development (Araral, 2013; Weible and Sabatier, 2018). However, considerably less attention has been given to policy concepts that quickly gain traction due to the apparent absence of disagreement in contexts where discord could be expected (however see Myerson and Rydin, 1996). Allocating greater attention to the role of meaning making can provide clarity of understanding on how problematic policy ambiguity is resolved in these circumstances. While variously applied, interpretive policy analysis emphasises the importance of such meaning making in the policy process. Here, significance is placed on the symbolic role of language, acts and objects as carriers of meaning. It is these meanings that constitute, consolidate and manifest the rationalities underpinning a new policy approach, and which help give it traction within the activities of a practitioner community, such as that of planning.

References

Allen, G. 2000. *Intertextuality*. Oxford, England, U.K., Routledge.

Ameel, L. 2020. *The Narrative Turn in Urban Planning: Plotting the Helsinki Waterfront*. Abingdon, England, U.K., Routledge.

Ansell, C. and Torfing, J. 2016. *Handbook on Theories of Governance*. Cheltenham, England, U.K., Edward Elgar Publishing.

Araral, E. 2013. *Routledge Handbook of Public Policy*. Abingdon, England, U. K., Routledge.

Aronowitz, S. 1988. *Science as Power: Discourse and Ideology in Modern Society*. Minneapolis, Minnesota, U.S.A., University of Minnesota Press.

Barthes, R. 2009[1957]. *Mythologies*. London, England, U.K., Vintage.

Beardsley, M. C. 1981[1958]. *Aesthetics: Problems in the Philosophy of Criticism*. Indianapolis, Indiana, U.S.A., Hacket Publishing Company.

Béland, D. 2019. *How Ideas and Institutions Shape the Politics of Public Policy*. Cambridge, England, U.K., Cambridge University Press.

Berger, P. and Luckmann, T. 1966. *The Social Construction of Reality: A Treatise on the Sociology of Knowledge*. London, England, U.K., Penguin Books.

Birkland, T. A. 2015. *Introduction to the Policy Process*. London, England, U.K., M.E. Sharpe.

Black, M. 1962. *Models and Metaphors*. Ithaca, New York, U.S.A., Cornell University Press.

Blacksell, M. 2006. *Political Geography*. Oxford, England, U.K., Routledge.

Bobrow, D. B. and Dryzek, J. S. 1987. *Policy Analysis by Design*. Pittsburgh, Pennsylvania, U.S.A., University of Pittsburgh Press.

Bogenschneider, K. and Corbett, T. 2021. *Evidence-Based Policymaking: Envisioning a New Era of Theory, Research, and Practice*. Abingdon, England, U.K., Routledge.

Bottici, C. 2007. *A Philosophy of Political Myth*. Cambridge, England, U.K., Cambridge University Press.

Bowker, G. C. and Leigh-Star, S. 1999. *Sorting Things Out: Classification and its Consequences*. Cambridge, Massachusetts, U.S.A., MIT Press.

Boyd, R. 1993. Metaphor and theory change: what is 'metaphor' a metaphor for? In: Ortony, A. (ed.) *Metaphor and Thought*. Cambridge, England, U.K., Cambridge University Press.

Burgess, J., Clark, J. and Harrison, C. M. 2000. Knowledges in action: an actor network analysis of a wetland agri-environment scheme. *Ecological Economics*, 35, 119–132.

Burke, K. 1966. *Language as Symbolic Action*. Berkley, California, U.S.A., University of California Press.

Burke, K. 1973. *The Philosophy of Literary Form*. Berkeley, California, U.S.A., University of California Press.

Cairney, P. 2019. *Understanding Public Policy: Theories and Issues*. London, England, U.K., Bloomsbury Publishing.

Chandler, D. 2007. *Semiotics: The Basics*. London, England, U.K., Routledge.

Charteris-Black, J. 2009. Metaphor and political communication. In: Musolff, A. and Zinken, J. (eds.) *Metaphor and Discourse*. Basingstoke, England, U.K., Palgrave MacMillan.

Clark, W. C. and Majone, G. 1985. The critical appraisal of scientific inquiries with policy implications. *Science, Technology and Human Values*, 10, 6–19.

Clemons, R. S. and McBeth, M. K. 2020. *Public Policy Praxis: A Case Approach for Understanding Policy and Analysis*. Abingdon, England, U.K., Routledge.

Colebatch, H. K. 2009. *Policy*. Berkshire, England, U.K., Open University Press.

Collins, A., Cowell, R. and Flynn, A. 2009. Evaluation and environmental governance: the institutionalisation of ecological footprinting. *Environment and Planning A*, 41, 1707–1725.

Compston, H. 2009. *Policy Networks and Policy Change: Putting Policy Network Theory to the Test*. Basingstoke, UK, Palgrave MacMillan.

Corner, J. 1999. The agency of mapping: speculation, critique and invention. In: Cosgrove, D. (ed.) *Mappings*. London, England, U.K., Reaktion Books.

Cosgrove, D. 1999. Introduction: mapping meaning. In: Cosgrove, D. (ed.) *Mappings*. London, England, U.K., Reaktion Books.

Couch, C. 2017. *Urban Planning: An Introduction*. London, England, U.K., Bloomsbury Publishing.

Crampton, J. W. 2003. *The Political Mapping of Cyberspace*. Edinburgh, Scotland, U.K., Edinburgh University Press.

Crowley, E. 2006. *Land Matters: Power Struggles in Rural Ireland*. Dublin, Ireland, The Lilliput Press.

Daniels, S., Delyser, D., Entrikin, N. J. and Richardson, D. 2011. Introduction: envisioning landscapes, making worlds. In: Daniels, S., Delyser, D., Entrikin, N. J. and Richardson, D. (eds.) *Envisioning Landscapes, Making Worlds: Geography and the Humanities*. Oxford, England, U.K., Routledge.

DeLeon, P. and Martell, C. R. 2006. The policy sciences: past, present and future. In: Peters, G. B. and Pierre, J. (eds.) *Handbook of Public Policy*. London, England, U.K., SAGE Publications Ltd.

Delmas, M. 2002. The diffusion of environmental management standards in Europe and the United States: an institutional perspective. *Policy Sciences*, 35, 91–119.

Dillon, S. and Craig, C. 2021. *Storylistening: Narrative Evidence and Public Reasoning*. Abingdon, England, U.K., Routledge.

Dodge, M., Kitchin, R. and Perkins, C. 2011. *Rethinking Maps: New Frontiers in Cartographic Theory*. Abingdon, England, U.K., Routledge.

Dorling, D. and Fairbairn, D. 1997. *Mapping: Ways of Representing the World*. Essex, England, U.K., Pearson Education Ltd.

Dreyfus, H. L. and Rabinow, P. 1982. *Michel Foucault: Beyond Structuralism and Hermeneutics*. Brighton, England, U.K., Harvester.

Dryzek, J. S. 2005. *The Politics of the Earth: Environmental Discourses*. Oxford, England, U.K., Oxford University Press.

Dryzek, J. S. 2013. *The Politics of the Earth: Environmental Discourses*. Oxford, England, U.K., Oxford University Press.

Eco, U. 1976. *A Theory of Semiotics*. Bloomington, Indiana, U.S.A., Indiana University Press.

Epstein, C. 2008. *The Power of Words in International Relations: Birth of an Anti-Whaling Discourse*. London, England, U.K., The MIT Press.

Evans, J. 2007. Wildlife corridors: an urban political ecology. *Local Environment*, 12, 129–152.

Feindt, P. and Oels, A. 2005. Does discourse matter? Analysis in environmental policy making. *Journal of Environmental Policy and Planning*, 7, 161–175.

Ferrari, E. and Rae, A. 2019. *GIS for Planning and the Built Environment: An Introduction to Spatial Analysis*. London, England, U.K., Bloomsbury Publishing.

Finch, J. 2021. *Literary Urban Studies and How to Practice It*. Abingdon, England, U.K., Routledge.

Fischer, F. 2003. *Reframing Public Policy: Discursive Politics and Deliberative Practices*. Oxford, England, U.K., Oxford University Press.

Flyvbjerg, B. 1998. *Rationality and Power: Democracy in Practice.* London, England, U.K., The University of Chicago Press Ltd.

Foucault, M. 1972[1969]. *The Archaeology of Knowledge.* London, England, U.K., Routledge.

Foucault, M. 1976. *The Will to Knowledge: The History of Sexuality: Volume 1.* London, England, U.K., Penguin Books Ltd.

Fry, B. R. and Raadschelders, J. C. N. 2008. *Mastering Public Administration: From Max Weber to Dwight Waldo.* Washington D.C., U.S.A., CQ Press.

Gane, N. 2004. *Max Weber and Postmodern Theory: Rationalization Versus Re-enchantment.* Basingstoke, England, U.K., Palgrave MacMillan.

Goatly, A. 1997. *The Language of Metaphor.* London, England, U.K., Routledge.

Goffman, E. 1979. Footing. *Semiotica,* 251–29.

Goffman, E. 1981. *Forms of Talk.* Oxford, England, U.K., Basil Blackwell.

Gold, J. R. and Revill, G. 2004. *Representing the Environment.* London, England, U.K., Routledge.

Goodin, R. E., Rein, M. and Moran, M. 2006. The public and its policies. In: Goodin, R. E., Rein, M. and Moran, M. (eds.) *The Oxford Handbook of Public Policy.* Oxford, England, U.K., Oxford University Press.

Goodwin, C. 2006. Interactive footing. In: Holt, E. and Clift, R. (eds.) *Reporting Talk: Reported Speech in Interaction.* Cambridge, England, U.K., Cambridge University Press.

Gottweis, H. 2003. Strategies of postcultural policy analysis. In: Hajer, M. and Wagenaar, H. (eds.) *Deliberative Policy Analysis: Understanding Governance in the Network Society.* Cambridge, U.K., Cambridge University Press.

Greenberg, G., Miller, J., Mohr, L. and Vladeck, B. 1977. Developing public policy theory: perspectives from empirical research. *American Political Science Review,* 71, 1532–1543.

Griggs, S. and Howarth, D. 2019. Discourse, policy and the environment: hegemony, statements and the analysis of UK airport expansion. *Journal of Environmental Policy & Planning,* 21, 464–478.

Hacking, I. 2002. *Historical Ontology.* Cambridge, Massachusetts, U.S.A., Harvard University Press.

Hajer, M. 1993. Discourse coalitions and institutionalization of practice: the case of acid rain in Britain. In: Fischer, F. and Forrest, J. (eds.) *The Argumentative Turn in Policy Analysis and Planning.* London, England, U.K., Duke University Press.

Hajer, M. 1995. *The Politics of Environmental Discourse: Ecological Modernisation and the Policy Process.* Oxford, England, U.K., Oxford University Press.

Hajer, M. 2005. Coalitions, practices, and meaning in environmental politics: from acid rain to BSE. In: Howarth, D. and Torfing, J. (eds.) *Discourse Theory in European Politics: Identity, Policy and Governance.* Basingstoke, England, U.K., Palgrave MacMillan.

Hajer, M. and Laws, D. 2006. Ordering through discourse. In: Moran, M., Rein, M. and Goodin, R. E. (eds.) *The Oxford Handbook of Public Policy.* Oxford, England, U.K., Oxford University Press.

Hajer, M. and Versteeg, W. 2005. A decade of discourse analysis of environmental politics: Achievements, challenges, perspectives. *Journal of Environmental Policy and Planning*, 7, 175–184.

Hall, T. 2019. *Town Planning: The Basics.* Abingdon, England, U.K., Routledge.

Halpern, C., Martinez, M., Van Breugel, I., Zittoun, P., Fischer, F. and Zahariadis, N. 2021. *The Political Formulation of Policy Solutions: Arguments, Arenas, and Coalitions.* Bristol, England, U.K., Policy Press.

Hannigan, J. 2014. *Environmental Sociology.* Abingdon, England, U.K., Routledge.

Harley, J. B. 1992. Deconstructing the map. In: Barnes, T. J. and Duncan, J. S. (eds.) *Writing Worlds: Discourse, Text and Metaphor in the Representation of Landscape.* London, England, U.K., Routledge.

Hart, C. 2008. Critical discourse analysis and metaphor: toward a theoretical framework. *Critcial Discourse Studies*, 5, 91–106.

Hausman, C. 2006. A revire of prominent theories of metaphor and metaphorical reference. *Semiotica*, 1, 213–230.

Hawkesworth, M. 2012. From policy frames to discursive politics: feminist approaches to development policy and planning in an era of globalization. In: Fischer, F. and Gottweis, H. (eds.) *The Argumentative Turn Revisited: Public Policy as Communicative Practice.* London, England, U.K., Duke University Press.

Hay, C. 2006. Constructivist Institutionalism. In: Rhodes, R. A. W., Binder, S. A. and Rockman, B. A. (eds.) *The Oxford Handbook of Political Institutions.* Oxford, England, U.K., Oxford University Press.

Heazle, M. and Kane, J. 2015. *Policy Legitimacy, Science and Political Authority: Knowledge and Action in Liberal Democracies.* Abingdon, England, U.K., Routledge.

Hébert, L. 2019. *An Introduction to Applied Semiotics: Tools for Text and Image Analysis.* Abingdon, England, U.K., Routledge.

Hill, M. 2009. *The Public Policy Process.* Essex, England, UK, Pearson Eduacation Limited.

Hill, M. and Varone, F. 2021. *The Public Policy Process.* Abingdon, England, U.K., Routledge.

Hogan, J. and Howlett, M. 2015. *Policy Paradigms in Theory and Practice: Discourses, Ideas and Anomalies in Public Policy Dynamics.* London, England, U.K., Palgrave Macmillan UK.

Howlett, C. 2010. Flogging a dead horse? Neo-Marxism and indigenous mining negotiations. *Australian Journal of Political Science*, 45, 457–474.

Howlett, M., Ramesh, M. and Perl, A. 2009. *Studying Public Policy: Policy Cycles & Policy Subsystems.* Ontario, Canada, Oxford University Press.

In't Veld, R. J. 2009. Willingly and knowingly: the roles of knowledge about nature and the environment in policy processes. In: In't Veld, R. J. (ed.) *RMNO-series Preliminary Studies and Background Studies (Number V.15e).* The Haag, The Netherlands, RMNO (Advisory Council for Spatial Planning, Nature and the Environment).

Isaksen, A. and Jakobsen, S. 2019. *Path Dependence and Regional Economic Renewal*. Abingdon, England, U.K., Routledge.

Johnstone, B. 2018. *Discourse Analysis*. Hoboken, New Jersey, U.S.A., Wiley.

Jones, M., Shanahan, E. and McBeth, M. 2014. *The Science of Stories: Applications of the Narrative Policy Framework in Public Policy Analysis*. New York, U.S.A., Palgrave Macmillan US.

Kaplan, T. 1993. Reading policy narratives: beginnings, middles and ends. In: Fischer, F. and Forrest, J. (eds.) *The Argumentative Turn in Policy Analysis and Planning*. London, England, U.K., Duke University Press.

Kenis, P. and Schneider, V. 1991. Policy networks and policy analysis: scruitinizing a new analytical toolbox. In: Marin, B. and Mayntz, R. (eds.) *Policy Networks: Empirical Evidence and Theoretical Considerations*. Frankfurt, Germany, Campus Verlag.

Kingdon, J. W. 1984. *Agendas, Alternatives and Public Politics*. New York City, New York, U.S.A., Harper Collins.

Kitchin, R., Perkins, C. and Dodge, M. (eds.) 2009. *Rethinking Maps: New Frontiers in Cartography*. Oxford, England, U.K., Routledge.

Knill, C. and Tosun, J. 2020. *Public Policy: A New Introduction*. London, England, U.K., Red Globe Press.

Knowles, M. and Moon, R. 2006. *Introducing Metaphor*. Oxford, England, U.K., Routledge.

Kövecses, Z. 2002. *Metaphor: A Practical Introduction*. Oxford, England, U.K., Oxford University Press.

Lakoff, G. and Johnson, M. 1980. *Metaphors We Live By*. London, England, U.K., University of Chicago Press.

Latour, B. 2005. *Reassembling the Social: An Introduction to Actor Network Theory*. Oxford, England, U.K., Oxford University Press.

Lejano, R., Ingram, M., Ingram, H., Kraft, M. E. and Kamieniecki, S. 2013. *The Power of Narrative in Environmental Networks*. Cambridge, Massachusetts, U.S.A., MIT Press.

Lemke, J. 1998. Multiplying meaning: visual and verbal semiotics in scientific text. *Reading Science: Critical and Functional Perspectives on Discourses of Science*. New York City, New York, U.S.A., Routledge.

Lennon, M. and Tubridy, F. 2022. 'Time'as a focus for planning research: exploring temporalities of coastal change. *Journal of Environmental Policy & Planning*, Online First: doi:10.1080/1523908X.2022.2122420.

Lennon, M. and Waldron, R. 2019. De-democratising the Irish planning system. *European Planning Studies*, 28, 1607–1625.

Lindblom, C. E. and Woodhouse, E. J. 1992. *The Policy-Making Process*. Essex, England, U.K., Pearson.

Lyon, G. 2000. Philosophical perspectives on metaphor. *Language Science*, 22, 137–153.

MacEachren, A. M. 1995. *How Maps Work: Representations, Visulaizations and Design*. New York, U.S.A., Guildford Press.

McNay, L. 1994. *Foucault: A Critical Introduction*. Cambridge, England, U.K., Polity Press.

Meadows, D. H., Randers, J. and Behrens, W. H. I. 1972. *The Limits to Growth*. New York, U.S.A., Universe Books.

Monmonier, M. 1991. *How To Lie With Maps*. Chicago, U.S.A., University of Chicago Press.

Moran, R. 1995. Metaphor. In: Wright, C. & Hale, B. (eds.) *Companion to the Philosophy of Language*. Oxford, England, U.K., Blackwell Publishing.

Moran, M., Rein, M. and Goodin, R. E. (eds.) 2009. *The Oxford Handbook of Political Science*. Oxford, England, U.K., Oxford University Press.

Murdoch, J. 2000. Networks - a new paradigm of rural development? *Journal of Rural Studies*, 16, 407–419.

Musolff, A. 2016. *Political Metaphor Analysis: Discourse and Scenarios*. London, England, U.K., Bloomsbury Publishing.

Myerson, G. and Rydin, Y. 1996. *The Language of Environment: A New Rhetoric*. London, England, U.K., UCL Press.

Nowlin, M. C. 2011. Theories of the policy process: state of the research and emerging trends. *Policy Studies Journal*, 39, 41–60.

O'Dwyer, B. and O'Sullivan, N. 2015. The structuration of issue-based fields: social accountability, social movements and the Equator Principles issues-based field. *Accounting, Organizations and Society*, 43, 33–55.

O'Sullivan, E. 2003. Marxism, the state and homelessness in Ireland. In: Ashead, M. and Millar, M. (eds.) *Public Administration and Public Policy in Ireland: theory and methods*. Oxford, England, U.K.: Routledge.

Ockwell, D. and Rydin, Y. 2006. Conflicting discourses of knowledge: understanding the policy adoption of pro-burning knowledge claims in Cape York Peninsula, Australia. *Environmental Politics*, 15, 379–398.

Orr, M. 2003. *Intertextuality*. Cambridge, England, U.K., Polity Press.

Ortony, A. (ed.) 1993. *Metaphor and Thought*. Cambridge, England, U.K., Cambridge University Press.

Owens, S., Rayner, T. and Bina, O. 2004. New agendas for appraisal: reflections on theory, practice, and research. *Environment and Planning A*, 36, 1943–1959.

Paltridge, B. 2006. *Discourse Analysis*. London, England, U.K., Continuum International Publishing Group.

Peters, B. G. 2021. *Advanced Introduction to Public Policy*. Cheltenham, England, U.K., Edward Elgar Publishing.

Phelps, N. A. 2021. *The Urban Planning Imagination: A Critical International Introduction*. Cambridge, England, U.K., Polity Press.

Pickles, J. 2004. *A History of Spaces: Cartogaphic Reason, Mapping and the Geo-Coded World*. Oxford, England, U.K., Routledge.

Potter, J. 1996. *Representing Reality*. London, England, U.K., SAGE Publications Ltd.

Rayner, J. 2009. Understanding policy change as a historical problem. *Journal of Comparative Policy Analysis: Research and Practice*, 11, 83–96.

Rhodes, R. A. W. 1997. *Understanding Governance: Policy Networks, Governance, Reflexivity, and Accountability.* Buckingham, England, U.K., Open University Press.

Rhodes, R. A. W. 2006. Policy network analysis. In: Moran, M., Rein, M. and Goodin, R. E. (eds.) *The Oxford Handbook of Public Policy.* Oxford, England, U.K.,Oxford University Press.

Rhodes, R. A. W. 2018. *Narrative Policy Analysis: Cases in Decentred Policy.* London, England, U.K., Palgrave Macmillan.

Richards, I. A. 1965[1936]. *The Philosophy of Rhetoric.* Oxford, England, U. K., Oxford University Press.

Richardson, T. 1996. Foucauldian discourse: power and truth in urban and regional policy making. *European Planning Studies*, 4, 279–292.

Ricoeur, P. 2002[1975]. *The Rule of Metaphor.* Oxford, England, U.K., Routledge.

Roe, E. 1994. *Narrative Policy Analysis: Theory and Practice.* London, England, U.K., Duke University Press.

Runhaar, H. 2009. Putting SEA in context: a discourse perspective on how SEA contributes to decision-making. *Environmental Impact Assessment Review*, 29, 200–209.

Rydin, Y. 2003. *Conflict, Consensus, and Rationality in Environmental Planning: An Institutional Discourse Approach.* Oxford, England, U.K., Oxford University Press.

Rydin, Y. 2007. Re-examining the role of knowledge within planning theory. *Planning Theory*, 6, 52–68.

Rydin, Y. and Tate, L. 2016. *Actor Networks of Planning: Exploring the Influence of Actor Network Theory.* Abingdon, England, U.K., Routledge.

Sabatier, P. A. 2007. The need for better theories. In: Sabatier, P. A. (ed.) *Theories of the Policy Process.* Boulder, Colorado, U.S.A.., Westview Press.

Sanders, E. 2006. Historical institutionalism. In: Rhodes, R. A. W., Binder, S. A. and Rockman, B. A. (eds.) *Political Institutions.* Oxford, England, U.K.: Oxford University Press.

Sanderson, I. 2002. Making sense of 'what works': evidence based policy making as instrumental rationality. *Public Policy and Administration*, 17, 61–75.

Schiappa, E. 2003. *Defining Reality.* Edwardsville, Ilinois, U.S.A., Southern Illinois University Press.

Schmidt, V. A. 2011. Speaking of change: why discourse is key to the dynamics of policy transformation. *Critical Policy Studies*, 5, 106–126.

Schön, D. A. 1993. Generative metaphor: a perspective on problem-setting in social policy. In: Ortony, A. (ed.) *Metaphor and Thought.* Cambridge, England, U.K.: Cambridge University Press.

Scott, W. R. 2008. *Institutes and Organisations: Ideas and Interests,* London, England, U.K., Sage Publications.

Semino, E. 2008. *Metaphor in Discourse.* Cambridge, England, U.K., Cambridge University Press.

Semino, E. and Demjén, Z. 2016. *The Routledge Handbook of Metaphor and Language.* Abingdon, England, U.K., Taylor & Francis.

Shapiro, M. 1981. *Language and Political Understanding.* New Haven, Connecticut, U.S.A., Yale University Press.

Sheridan, A. 1980. *Michel Foucault: The Will to Truth.* Oxford, England, U. K., Tavistock Publications Ltd.

Smart, B. 1985. *Michel Foucault.* Cornwall, U.K., Taylor & Francis.

Smith, K. B. and Larimer, C. W. 2009. *The Public Policy Theory Primer.* Boulder, Colorado, U.S.A., Westview Press.

Steffek, J. 2003. The legitimation of international governance: a discourse approach. *European Journal of International Relations,* 9, 249–275.

Steffek, J. 2009. Discursive legitimation in environmental governance. *Forest Policy and Economics,* 11, 313–318.

Stone, D. 1989. Causal stories and the formation of policy agendas. *Political Science Quarterly,* 104, 281–300.

Stone, D. 1997. *Policy paradox: The Art of Political Decision Making.* New York, U.S.A., W.W. Norton.

Swaffield, S. 1998. Contextual meanings in policy discourse: a case study of language use concerning resource policy in the New Zealand high country. *Policy Sciences,* 31, 199–224.

Swain, C. and Tait, M. 2007. The crisis of trust and planning. *Planning Theory and Practice,* 8, 229–247.

Thalen, K. 1999. Historical institutionalism in comparative politics. *American Review of Political Science,* 2, 369–404.

Throgmorton, J. A. 1993. Survey research as rhetorical trope: electric power planning arguements in Chicago. In: Fischer, F. and Forester, J. (eds.) *The Argumentative Turn in Policy Analysis and Planning.* London, England, U. K., Duke University Press.

Tolson, A. 2006. *Media Talk: Spoken Discourse on TV and Radio.* Edinburgh, Scotland, U.K., Edinburgh University Press.

Torfing, J. 2005. Discourse theory: achievements, arguments and challenges. In: Howarth, D. and Torfing, J. (eds.) *Discourse Theory in European Politics: Identity, Policy and Governance.* Basingstoke, England, U.K., Palgrave MacMillan.

Wagenaar, H. 2011. *Meaning in Action: Interpretation and Dialogue in Policy Analysis.* Armonk, New York, U.S.A., M.E. Sharpe.

Wagenaar, H. and Cook, S. D. N. 2003. Understanding policy practices: action, dialectic and deliberation in policy analysis. In: Hajer, M. and Wagenaar, H. (eds.) *Deliberative Policy Analysis: Understanding Governance in the Network Society.* Cambridge, England, U.K., Cambridge University Press.

Waldron, R. 2019. Financialization, urban governance and the planning system: utilizing 'development viability' as a policy narrative for the liberalization of Ireland's post-crash planning system. *International Journal of Urban and Regional Research,* 43, 685–704.

Weber, M. 1922. *The Theory of Social and Economic Organisation.* New York, U.S.A., The Free Press.

Weible, C. M. and Sabatier, P. A. 2018. *Theories of the Policy Process.* Abingdon, England, U.K., Routledge.

Weible, C. M. and Workman, S. 2022. *Methods of the Policy Process.* New York, U.S.A., Routledge.

Wilson, M. W. 2017. *New Lines: Critical GIS and the Trouble of the Map.* Minneapolis, Minnesota, U.S.A., University of Minnesota Press.

Wood, D. 1992. *The Power of Maps.* New York, U.S.A., The Guildford Press.

Wu, X., Ramesh, M., Howlett, M. and Fritzen, S. A. 2017. *The Public Policy Primer: Managing the Policy Process.* Abingdon, England, U.K., Routledge.

Yanow, D. 1996. *How Does a Policy Mean? Interpreting Policy and Organisational Actions.* Washington DC, U.S.A., Georgetown University Press.

Yanow, D. 2000. *Conducting Interpretive Policy Analysis.* London, England, U.K., Sage Publications.

Yanow, D. 2002. *Making American 'Race' and 'Ethnicity': category failures in policy and administrative practices.* Armonk, New York, U.S.A., M.E.Sharpe.

Zahariadis, N. 2003. *Ambiguity and Choice in Public Policy: Political Decision Making in Modern Democracies.* Washington DC, U.S.A., Georgetown Press.

2 Case Study Summary

What is Green Infrastructure?

Green infrastructure (GI) is an intuitively understood concept that is nevertheless difficult to universally define. This is because it is employed in different ways in different contexts: in some jurisdictions it is most often conceived as a tool for evaluating specific development proposals, while in others it is mainly employed in the formulation of strategic spatial frameworks (Matsler et al., 2021). Much of this variation results from how the roots of the concept straddle a number of disciplines, with the relative weight given to each of these roots reflected in the focus of GI initiatives in different contexts (Lennon, 2019; de Oliveira and Mell, 2019). For example, in North America, GI is most often focused on harnessing natural processes for sustainable drainage management (Brears, 2018; Rouse and Bunster-Ossa, 2013). Drainage is also to the fore in China (Chan et al., 2018; Qiao et al., 2020; Yin et al., 2021), although urban heat island mitigation is given more representation (Shao and Kim, 2022). On the African continent, drainage is also of significant concern, although here the array of climates, environments and socio-economic contexts means that GI is more associated with a drive to address the challenges of rapid urbanisation in financially constrained conditions than with a single shared environmental challenge (Abbot, 2012).

In Continental Europe, (non-China) Asia and Oceania, emphasis varies within and between countries, with a focus shifting from biodiversity conservation to urban heat island mitigation and/or drainage management depending on the context (Di Marino and Lapintie, 2018; Mell, 2016; Lund et al., 2019; Dhyani et al., 2022). While in the UK and Ireland, attention is given to all these issues, but usually with additional weight given to active transport, aesthetic enhancement and recreation space provision (Lennon et al., 2016; Washbourne and Wansbury, 2023). Thus, rather than deriving a generalised and universally applicable

DOI: 10.4324/9781003413608-3

definition, understanding 'what' GI signifies involves appreciating the context in which it emerges and the purposes it is seen to serve by those employing it. Accordingly, investigating the emergence of GI in a particular context proffers the potential for insight into the role of meaning making in the planning policy process.

Nonetheless, the GI concept shares a number of outline commonalities across the many contexts in which it enjoys representation. These largely centre on the idea of GI as a form of 'infrastructure' that underpins the functioning of society. In this sense, GI is widely perceived as something that can be planned, designed, delivered and implemented, as has been done for centuries, with conventional 'grey' infrastructure (pipes, roads, cables). In this manner, GI can be conceived as a 'policy approach', wherein a policy approach is understood as a set of propositions guiding policy formulation and implementation that reflect rationalities through the implicit commonalities of policies comprising a broadly shared perspective on how a particular set of problems should be addressed.

Although those advocating a GI approach in planning claim antecedents in established practices (Mell, 2016; Lennon, 2018; Duvall et al., 2018), the emergence and swift ascent to prominence of specific 'green infrastructure' policy discourses is only two decades old. Hence, tracing the emergence of this approach proffers potential for a comprehensive understanding of the role played by meaning making in resolving issues of problematic policy ambiguity. Indeed, key to understanding the GI approach is how it proposes a new perspective on remedying persistent problems plagued by policy ambiguity, such as how to effectively facilitate biodiversity conservation in planning contexts where pressing issues such as drainage management, transport infrastructure provision and the delivery of public recreation space compete for attention on the decision-making agenda. Consequently, appreciating GI as a new policy approach entails understanding it as the emergence of a new, or at least a different, perception of how a set of landuse planning problems should be attended to. Although 'change' to existing policy may constitute an important element in this process through its aggregated affects and effects at particular levels of the policy hierarchy, the emergence of the GI approach has permeated multiple levels of the policy hierarchy in numerous contexts, thereby broadly influencing the governance of a particular set of issues. In this sense, appreciating the emergence of GI involves attention to 'the principles of governance' in relation to these issues. 'Governance' is here understood as the activity of establishing, affirming or changing the principles employed to mediate current or potential conflicts in administering contemporary

conditions or planning for the future. As such, the principles of governance are manifested in the meanings that are given expression both in the
policy proposals forwarded by the promoters of GI and those who subscribe to it as a policy approach. An understanding of the role of
interpretation and representation in the policy process is thereby
necessary. Particular cognisance must thus be shown to how the 'facts'
of a situation are perceived through the accounts given of them by
interpreting agents (Fischer, 2003; Rabinow and Sullivan, 1992; Schön
and Rein, 1994; Yanow, 2007). This is not to deny the concept of
'objectivity' per se, but rather to acknowledge that 'the styles of reasoning that we employ determine what counts as objectivity' (Hacking,
2002, 160–161).

The case of GI's emergence in Ireland provides a particularly fitting
location to examine the role of meaning making in the planning policy
process. Between 1995 and 2008, Ireland experienced considerable
economic, demographic and urban growth (Kirby, 2010; O'Hagan and
Newman, 2011). During this period, landuse governance struggled to
negotiate the complex planning and environmental policy issues associated with unprecedented pressures for urban and infrastructural
development (Davies, 2008; McCann, 2011; Yarwood, 2006). While
growth rates significantly reduced following the financial crash of 2008
and the subsequent Great Recession, policy issues associated with over
a decade of intense development demands remained (Kitchin et al.,
2012). Keeping pace with such growth, and subsequently addressing its
consequences, preoccupied planning policy activity in Ireland during
and after the financial crash. It is against this backdrop that new policy
approaches such as GI were sought to solve multiple complex and
pressing landuse governance issues. Specifically, in November 2008,
Fingal County Council organised a 'Green Infrastructure Conference' in
Malahide, North County Dublin, Ireland. Prior to this, mention of GI in
Irish landuse planning advocacy and guidance documentation had been
limited (Tubridy and O Riain, 2002; UCD et al., 2008). However, following this conference, reference to GI in such documentation increased
considerably over the subsequent three years. Indeed, in just three years –
by November 2011 – the GI concept was cited in Irish statutory planning
policy at national, regional and local levels, while also enjoying significant
representation in many non-statutory planning policy and advocacy
documents.

A review of the concept's history in Ireland reveals that interpretations of GI's meaning evolved and broadened significantly since it was
first cited. Nevertheless, close scrutiny of how GI's meaning was construed reveals that the diverse interpretations of GI were united by a

common thread, namely the concept's applicability to a spectrum of broadly conceived 'green spaces' and its specification as 'infrastructure': something of necessity that could be planned and delivered in remedying the existing or predicted problems of development. It is in this sense of manifesting a different, yet widely-shared perspective on how a particular set of problems should be addressed that the GI concept can be considered a new 'policy approach' to the planning and management of broadly-conceived 'green spaces'.

Ireland's relatively small population of just 4.6 million (CSO, 2011) during the period of GI's emergence between November 2008 and November 2011, and the limited number of planning authorities within the country means that tracing the undisputed emergence of the GI approach in Ireland is facilitated by the restricted number of actors concerned with its advocacy in planning policy formulation. This restricted administrative, spatial and temporal context thus renders it feasible to comprehensively chart the role played by meaning making in the rapid emergence and adoption of the GI planning approach in Ireland. This involves asking *why* did the concept emerge in Irish planning policy debates, *what* it meant, and *how* did these meanings evolve? No sooner have such questions being broached than an additional series of closely associated queries emerge as to *who* advanced the concept and *why* did they chose to do so, *how* was the concept advocated, and *why* did it attract so much attention? The case study that follows is used to identify and explain the role of meaning making in the policy process by addressing these questions. To achieve this, the case study draws on an extensive documentary analysis of both international and Irish planning policy related material. Complementing this is the analysis of semi-structured interviews with 52 interviewees conducted between April and September in 2011. A purposeful and snowball sampling method was used to identify the interviewees, who emendated from a cross-section of the public, quasi-autonomous non-governmental organisation (QUANGO), non-governmental organisation (NGO) and private sectors. Data from participant observation in two workshops in 2010 is used to supplement the documentary analysis and interviews. But before delving into a detailed interpretation of the data, it is first beneficial to outline the broad contours of the emergence and evolution of GI in Ireland.

The Emergence and Evolution of GI in Ireland

The story of the GI approach in Ireland is one in which the concept emerged and evolved from an ecologically centred 'networked' approach to conservation into a perspective increasingly focused on the planning of a broadly-encompassing conception of 'green spaces' for anthropocentric utility. This reorientation of GI's meaning increasingly sought to emphasise the services such areas provide in aiding physical, social and economic development. This evolving reconceptualisation of what GI signifies may be divided into three phases, namely: 2002–2007; 2008; and 2009–2011, each of which is summarised below.

First Phase: 2002–2007

The initial period of the concept's manifestation between 2002–2007 was characterised by a chronological sequence in the realignment of the networked approaches to green spaces. This succession of antecedent discourses to GI commenced with the appearance in 2002 of an 'ecological network' approach that foregrounded the conservation of habitats. The popularity of this approach persisted until 2005 when it was subsequently overtaken by a 'green network' concept, which with greater standing in statutory planning guidance, assumed the compatibility of multifunctional landuses in the provision of open space planning and the management of natural heritage. The third period, discernible during the 2005–2007 phase, not only reflects an interpretation of multifunctional compatibilities but further extends this discourse to advocate a 'green network' approach as one in which 'using' nature is seen as a means to protect it.

Second Phase: 2008

By early 2008, discourses surrounding planning for biodiversity had broadened to include a wide remit of uses in addition to the already popular assumption of biodiversity protection and recreational landuse compatibilities. The emergence of an ecosystems services discourse further repositioned perspectives on planning for biodiversity away from concepts focused on the intrinsic value of nature towards those concerned with the anthropocentric instrumental value of 'ecological assets' (DoEHLG, 2008). Whilst such an instrumental perspective appears to have dominated the planning literature during this period, there is evidence to suggest the persistence of interpretations that maintained a bias towards the intrinsic values of nature in planning policy proposals. The publication of the Green City Guidelines in September of the same year (UCD et al., 2008) observed the reintroduction into planning discourses of the term 'green infrastructure'. The Green Infrastructure Conference of November 2008 consolidated the reappearance of GI as a planning discourse and witnessed a number of interpretations of GI, although those with a specific 'planning' focus emphasised the anthropocentric utility value of broadly-conceived green spaces.

Third Phase: 2009–2011

The 2009–2011 period witnessed a considerable expansion in the interpretation of GI's spatial and functional applicability. Almost all spatial typographies, including brownfield sites (DCC, 2009) and cultural heritage locations (DRA and MERA, 2010), were interpreted as constituent elements of GI. Simultaneously, the functions of GI were expanded and coupled to discourses on economic development which stretched beyond the planning arena and into contemporary themes in wider politics and society (Clabby, 2009; Comhar, 2010a; Comhar, 2010b).[1] Reinforcing this association, 2010 witnessed an escalating monetarisation of biodiversity issues by way of reference to ecosystems services. Here, GI became increasingly fashioned as a planning mechanism underpinned by a sound economic rationale (Comhar, 2010b; DoEHLG, 2010). This year also saw the emergence of a movement to foster a quantitatively-based cartographic foundation for the formulation and implementation of GI planning. Furthermore, evident in late 2010 through to 2011 was the increasing prominence of professional institutes in advocating GI. Throughout this period, varying interpretations of what GI entails continued to flourish. By autumn 2011, conceptions of GI had moved well beyond 'networked' spatial arrangements. GI was now increasingly applied as a label indicating the society-servicing functions of all green spaces, be they connected or isolated, naturally occurring or human made. By the end of 2011, GI had achieved endorsement in statutory documents at all levels of the planning hierarchy. However, with the exception of Galway City Council on Ireland's west coast, the most comprehensive representation of GI was in the Greater Dublin Area, and more specifically within the local authorities comprising the Dublin metropolitan region[2].

In summary, what GI means developed from a 'networked' approach for the conservation of habitats to a multifunctional and anthropocentric 'ecosystems services' approach to green space planning. The following chapters examine 'why' and 'how' this transformation occurred. These furnish a detailed illustration of the mechanics of 'Policy Entitlement'.

Notes

1 Works referenced throughout this book from the Comhar SDC website no longer have a working web link to the website referenced. This is because Comhar has been permanently closed since 2011 and its work integrated

into another government body in 2012. The Comhar website no longer exists.
2 Dublin City Council, Dun Laoghaire-Rathdown County Council, South Dublin County Council, and Fingal County Council.

References

Abbot, J. 2012. *Green Infrastructure for Sustainable Urban Design in Africa.* Oxford, England, U.K., Earthscan.

Brears, R. C. 2018. *Blue and Green Cities: The Role of Blue-Green Infrastructure in Managing Urban Water Resources.* London, England, U.K., Palgrave Macmillan UK.

Chan, F. K. S., Griffiths, J. A., Higgitt, D., Xu, S., Zhu, F., Tang, Y.-T., Xu, Y. and Thorne, C. R. 2018. "Sponge City" in China—a breakthrough of planning and flood risk management in the urban context. *Land Use Policy*, 76, 772–778.

Clabby, G. 2009. *Green Infrastructure: Critical Infrastructure for a Smart Economy* [Online]. Dublin, Ireland, Comhar SDC. [Accessed 12.11.10].

Comhar. 2010a. *Creating Green Infrastructure for Ireland: Enhancing Natural Capital for Human Well Being.* Dublin, Ireland, Comhar SDC.

Comhar. 2010b. *Workshop on The Economics of Ecosystems and Biodiversity.* Dublin, Ireland, Comhar SDC.

CSO. 2011. *2011 Census of Population: This is Ireland - Highlights from the Census, Part 1* [Online]. Dublin, Ireland, Central Statistics Office. Available: http://www.cso.ie/en/census/census2011reports/census2011thisisirelandpart1/ [Accessed 18 June 2012].

Davies, A. 2008. *The Geographies of Garbage Governance in Ireland: Interventions, Interactions and Outcomes.* Aldershot, England, U.K., Ashgate Publishing Limited.

DCC. 2009. *Draft Dublin City Development Plan 2011–2017.* Dublin, Ireland, Dublin City Council.

de Oliveira, F. L. and Mell, I. 2019. *Planning Cities with Nature: Theories, Strategies and Methods.* Cham, Switzerland, Springer International Publishing.

Dhyani, S., Basu, M., Santhanam, H. and Dasgupta, R. 2022. *Blue-Green Infrastructure Across Asian Countries: Improving Urban Resilience and Sustainability.* Singapore, Springer Nature Singapore.

Di Marino, M. and Lapintie, K. 2018. Exploring the concept of green infrastructure in urban landscape. Experiences from Italy, Canada and Finland. *Landscape Research, 43*, 139–149.

DoEHLG. 2008. *The Economic and Social Aspects of Biodiversity: Benefits and Costs of Biodiversity in Ireland.* Dublin, Ireland, Government of Ireland.

DoEHLG. 2010. *Draft National Biodiversity Plan 2010–2015.* Dublin, Ireland, Government of Ireland.

DRA and MERA. 2010. *Regional Planning Guidelines for the Greater Dublin Area 2010–2022.* Dublin, Ireland, Mid-East Regional Authority and Dublin Regional Authority.

Duvall, P., Lennon, M. and Scott, M. 2018. The 'natures' of planning: evolving conceptualizations of nature as expressed in urban planning theory and practice. *European Planning Studies*, 26, 480–501.

Fischer, F. 2003. *Reframing Public Policy: Discursive Politics and Deliberative Practices*. Oxford, U.K., Oxford University Press.

Hacking, I. 2002. *Historical Ontology*. Cambridge, Massachusetts, U.S.A., Harvard University Press.

Kirby, P. 2010. *Celtic Tiger in Collapse: Explaining the Weaknesses of the Irish Model*. Basingstoke, U.K., Palgrave MacMillian.

Kitchin, R., O'Callaghan, C., Boyle, M., Gleeson, J. and Keaveney, K. 2012. Placing neoliberalism: the rise and fall of Ireland's Celtic Tiger. *Environment and Planning A*, 44, 1302–1326.

Lennon, M. 2018. Grasping Green Infrastructure: an introduction to the theory and practice of a diverse environmental planning approach. In: Davoudi, S., Blanco, H., Cowell, R. and White, I. (eds.) *Routledge Companion to Environmental Planning and Sustainability*. London, England, U.K., Routledge.

Lennon, M., Scott, M., Collier, M. and Foley, K. 2016. The emergence of green infrastructure as promoting the centralisation of a landscape perspective in spatial planning—the case of Ireland. *Landscape Research*, 42, 146–163.

Lund, N. S. V., Borup, M., Madsen, H., Mark, O., Arnbjerg-Nielsen, K. and Mikkelsen, P. S. 2019. Integrated stormwater inflow control for sewers and green structures in urban landscapes. *Nature Sustainability*, 2, 1003–1010.

Matsler, A. M., Meerow, S., Mell, I. C. and Pavao-Zuckerman, M. A. 2021. A "green" chameleon: exploring the many disciplinary definitions, goals, and forms of "green infrastructure". *Landscape and Urban Planning*, 214, 104145.

McCann, G. 2011. *Ireland's Economic History: Crisis and Development in the North and South*. London, England, U.K., Pluto Press.

Mell, I. 2016. *Global Green Infrastructure: Lessons for Successful Policymaking, Investment and Management*. London, England, U.K., Taylor & Francis.

O'Hagan, J. and Newman, C. (eds.) 2011. *The Economy of Ireland: National and Sectoral Policy Issues*. Dublin, Ireland, Gill & Macmillan.

Qiao, X.-J., Liao, K.-H. and Randrup, T. B. 2020. Sustainable stormwater management: a qualitative case study of the Sponge Cities initiative in China. *Sustainable Cities and Society*, 53, 101963.

Rabinow, P. and Sullivan, W. M. 1992. The interpretive turn: a second look. In: Rabinow, P. and Sullivan, W. M. (eds.) *Interpretive Social Science: A Second Look*. Berkeley, California, U.S.A.: University of California Press.

Rouse, D. C. and Bunster-Ossa, I. F. 2013. *Green Infrastructure: A Landscape Approach*. Washington, D.C., U.S.A., American Planning Association.

Schön, D. A. and Rein, M. 1994. *Frame Reflection: Towards the Resolution of Intractable Policy Controversies*. New York, U.S.A., Basic Books.

Shao, H. and Kim, G. 2022. A comprehensive review of different types of green infrastructure to mitigate urban heat islands: progress, functions, and benefits. *Land*, 11, 1792.

Tubridy, M. and O Riain, G. 2002. *Preliminary Study of the Needs Associated with a National Ecological Network*. Wexford, Ireland, Environmental Protection Agency.

UCD, DLRCC, FCC and NATURA. 2008. *Green City Guidelines*. Dublin, Ireland, UCD Urban Institute.

Washbourne, C.-L. and Wansbury, C. (eds.) 2023. *ICE Manual of Blue-Green Infrastructure*. London, England, U.K., ICE Publishing.

Yanow, D. 2007. Interpretation in policy analysis: on methods and practice. *Critical Policy Studies*, 1, 110–122.

Yarwood, J. (ed.) 2006. *The Dublin-Belfast Development Corridor: Ireland's Mega-City Region?*Aldershot, Hampshire, England, Ashgate Publishing Limited.

Yin, D., Chen, Y., Jia, H., Wang, Q., Chen, Z., Xu, C., Li, Q., Wang, W., Yang, Y. and Fu, G. 2021. Sponge city practice in China: a review of construction, assessment, operational and maintenance. *Journal of Cleaner Production*, 280, 124963.

3 Narrative Production

Problematising

A Root Problem Narrative

The problem of habitat fragmentation had been identified as an issue requiring remedy within the first formal reference to 'green infrastructure' (GI) in an Irish policy context. This study commissioned by the Irish Environmental Protection Agency (EPA), equated GI with ecological networks noting that,

> [t]here is evidence for the effects of fragmentation on habitat and species connectivity in Ireland … A spatial solution to this problem can be elaborated under the broad title of 'ecological network' …
> (Tubridy and O Riain, 2002, 46)

The perception of habitat fragmentation as a problem necessitating redress by the planning system appears to have persisted in the absence of specific GI policy representation between 2003 and 2007. This was illustrated by means of reference to the importance of ecological connectivity in a variety of increasingly multifunctional 'networked' approaches to green space planning, including among others, 'Green Networks' (GCC, 2005), 'Green Chains' (DCC, 2005), and 'Ecological Networks' (FCC, 2005a, 2005b). The 'problem' of habitat fragmentation was once again noted in the second formal reference to GI in an Irish planning policy context in the Green City Guidelines (UCD et al., 2008). This was published in November 2008, six years after the first reference to GI in the EPA study (Tubridy and O Riain, 2002). Although by this time the 'networked' approach to biodiversity planning had substantially departed from that originally advocated in the EPA study, having now morphed into a largely anthropocentric

DOI: 10.4324/9781003413608-4

instrumental and multifunctional perspective on green space provision, the 2008 Green City Guidelines note that,

> [p]lanning for biodiversity must take the spatial requirements of species into consideration by providing sufficient habitat for them in a connected arrangement. A spatial overview at the landscape-scale is required to overcome existing fragmentation and prevent further depletion of connected features.
>
> (UCD et al. 2008, 15)

This focus on tackling habitat fragmentation was subsequently represented in presentations at workshops organised by Comhar SDC[1] (Comhar, 2010a, 2010b), and the published Comhar SDC study on 'Creating Green Infrastructure for Ireland' (Comhar, 2010c). Such a concern with the negative ecological consequences resulting from habitat fragmentation was echoed across a broad range of professional disciplines associated with the emergence and evolution of GI planning policy discourses in Ireland. This is lucidly exemplified by one local planning authority interviewee who observed that,

> [w]e have dots, at one point it wasn't dots, it was a complete you know, a landscape or an interconnected landscape but ... we've introduced fragmentation and now we have to actually plan connectivity where that didn't have to be planned before.
>
> (Interviewee B1)

The endurance of habitat fragmentation as a problematic discourse formerly grounding, and latterly intrinsic to, appeals for greater attention to green space connectivity in planning policy, suggests a 'root problem narrative' which formed a common concern threading through all discussions on 'networked' approaches to green space planning. Although biodiversity loss was conceived as directly resultant from such habitat fragmentation ('root problem narrative'), this discourse was conceived as nested within a wider narrative pivoting on impressions of a prevalent malaise in landuse governance. This was attributed to the perceived low profile of natural heritage issues in planning policy. Such an encompassing, but more nebulous predicament, had ambiguous foundations and consequently a less definable solution (Barry, 2007; Yearley, 2002). As such, this issue of low profile presented those concerned about it with problematic policy ambiguity. Nevertheless, given its perceived encompassing position, it was to this 'broader problem narrative' that efforts to address habitat fragmentation were directed.

A Broader Problem Narrative

The perceived low profile of natural heritage issues in landuse governance was succinctly expressed by Comhar when it pronounced that,

> [b]iodiversity continues to decline because its value is not reflected in decision making by business and Government.
>
> (Comhar, 2010c, 5)

This view was echoed by many of those interviewed, with some noting that this poor status was evidenced in normal professional practice. It is in this context that one senior local authority officer commented that,

> [m]y view, of the last twenty two years ... is that a lot of professionals take little heed of the natural environment. They do not see the consequence of what they do to be honest with you.
>
> (Interviewee B6)

Several of those interviewed observed how a directed bias in policy formulation practices marginalised certain planning related issues, as was noted by the conclusion that,

> I suppose the whole green side of things ... that space is always compiled as the left over space, you know, your left over, the stuff you haven't zoned.
>
> (Interviewee E4)

Such appraisals stimulated the emergence of countering discourses that sought to advance perspectives on the importance of these areas by emphasising their potential as multifunctional resources for the delivery of both societal and ecological benefits. This was conveyed by a much referenced Comhar Commentary,[2] which contended that,

> [t]he importance of developing and maintaining different types of infrastructure for future economic wellbeing is clear. Because of this, we carefully plan our road networks, our power supply networks, our telecoms networks. We invest heavily in these strategic assets and maintain them on an ongoing basis.
>
> In contrast, we think about and manage land and green space in a very different way. We see individual parcels rather than a connected network. We usually don't think about the many benefits

being provided to us, often free of charge. And we don't recognise that these networks of land and green space surrounding our towns and cities – and threaded through them – play a key role in sustaining environmental quality.

(Clabby, 2009)

In this sense, Comhar sought to stress the need to consider green spaces as essential to the economic, social and environmental sustainability of existing and future development. Most of those interviewed suggested that raising awareness of this crucial servicing function was required so that such areas, and the issues associated with them, were allocated greater weight in policy formulation. Devising a means by which to communicate the fundamental societal importance of green spaces was thus viewed as vital in facilitating the associated protection of ecosystem integrity and the prevention of habitat fragmentation. To many, this communication agenda entailed providing clarity and direction to the disparate planning practices associated with such areas. As noted by one senior planning official,

I suppose maybe there is a sense that maybe it needs greater prominence, you know, greater priority, greater focus and one of the ways in which you do that is to, you know, carve out a particular identity and conceptual framework for it and promote it on that basis.

(Interviewee C10)

It was this concern with attaining 'prominence' while concurrently establishing a 'particular identity and conceptual framework' that lay at the heart of the 'solution narrative' presented by those advocating a GI approach to planning policy formulation.

Solving

Seeing Green

For the majority of those interviewed, terming green spaces as 'GI' offered a solution to the problem of communicating the importance of these areas. An example of this labelling function is illustrated by the Regional Planning Guidelines for the Greater Dublin Area 2010–2022 when it states,

Green Infrastructure (GI) is a generic term encompassing the protection, management and enhancement of urban, peri-urban and rural environmental resources (natural and managed) through the identification and provision of multifunctional and interconnected green spaces and provides an opportunity to reassess the manner in which we manage and use our green spaces.

(DRA and MERA, 2010, 159)

As such, employing the term 'GI' was conceived as a means by which to move perceptions regarding green spaces 'away from this idea … that like land that isn't being developed is just sitting there doing nothing … it isn't just sitting there doing nothing, it's doing something' (Interviewee B20). This shift in perspectives on green spaces from 'doing nothing' to services provision was widely regarded by those interviewed as attributable to the labelling of green spaces as 'infrastructure'. As stated by one interviewee, but repeated by many others,

> I think that's a fairly, very powerful concept, you know to most people. They think infrastructure is something useful, so you're kind of making people think mmm, there's some use in this green stuff you know. It makes you aware of that you know, makes you think about that.
>
> (Interviewee A5)

The recognition of 'use in this green stuff' consequent on terming it 'infrastructure' was perceived by most of those interviewed as raising the profile of green spaces in planning. This was conveyed by an advocate of GI when asserting,

> I'm a typical frustrated landscape architect, always feeling that landscape or open space is left behind as an afterthought to the planning system and at strategic level and on projects for that matter and I think green infrastructure as a concept brings it to the centre or to the forefront of planning as I feel it should be.
>
> (Interviewee A2)

Such perceived success in foregrounding green space considerations in planning was credited by many of those interviewed to the placing of green space issues on an equal policy standing with that of other competing issues in the plan making process. Again, this success was widely credited to labelling such areas as 'GI'. This is expressed in the opinion that,

> [o]ften in planning terms I suppose … the green space probably gets overlooked a little bit in terms of land use, zoning, and plan

making processes ... the idea of about where we put residential, where we put our employment, might be more headline issues than, than where we put our green space sometimes. So, I think that the concept can really, it can broaden the integration of these spaces and what they're actually used for and how they relate to the planning system. So I think it's a good mechanism and a good tool to sort of mainstream the whole idea of that whole issue.

(Interviewee B2)

The 'integration' of such a mainstreamed concept entailed a transformation in the perception of green spaces from 'the left over space ... the stuff you haven't zoned' into areas requiring consideration at early stages in the plan making process. As declared by one planner,

[i]t's taking a more proactive approach to the creation of green spaces and the design of green spaces to make sure that ... it's plan led in some way and not something that's accidental, that just falls out of a plan when all the hard construction is put in place.

(Interviewee C5)

Seeing green spaces, and the issues associated with them, as items requiring proactive consideration in the plan formulation process was largely attributed to the role of the term 'GI' in communicating the value of such areas. GI was thus viewed as conferring on green space planning issues an increased degree of 'discursive weight' in discussions concerning the appropriate use of land.

Discursive Weight

A number of interviewees expressed the view that garnering increased weight of consideration for green space planning, and their associated issues, amid a competing array of concerns in landuse policy formulation involves placing greater emphasis on anthropocentric utility in deliberations about such areas. To most of those interviewed, the term 'GI' facilitates this manoeuvre, an opinion lucidly articulated in Comhar's report advocating for a GI approach to planning,

[t]here is general dissatisfaction with the mechanisms currently available to input information on biodiversity to spatial plans. Respondents, to whom the concept was introduced directly for the first time, considered that the concept of Green Infrastructure and mechanism of Green Infrastructure planning will be more attractive

than ecological networks because of the clearer focus on benefits to people.

(Comhar, 2010c, 22)

This assessment of the need to accentuate the 'benefits to people' provided by GI planning emerged as a frequently-referenced issue during interviewing. Specifically, the majority of interviewees from planning authorities expressed the view that the anthropocentric focus of GI provided a more effective means of gaining attention for green space issues than efforts centred on the advocacy of ecological networks. Such a belief was conveyed by one such local authority planner when noting,

[i]t's a better descriptive term and it's a more proactive term where you're actually trying to create something or, whereas you know, 'ecological networks' is very, I mean it was the sort of, the buzz phrase of you know, ten years ago. I don't think it ever really worked, certainly not in this country, there doesn't seem to be that much, not in my experience, there didn't seem to be that much done with it.

(Interviewee B10)

To most of those interviewed, this perception of proactivity and creativity is effected by the inclusion of the word 'infrastructure' in the term 'GI'. Indeed, for many interviewees the word 'infrastructure' was identified as a means of enabling communication between those engaged in the production of planning policy and those advocating the importance of green space planning. As concluded by one consultant,

[t]he use of the term 'infrastructure' is quite useful, you know, local authority planners and so on get it, and they can sell it a lot better ... It certainly is a big improvement on 'ecological network', which doesn't get them, doesn't grasp them as much I think.

(Interviewee A4)

The perceived benefit of employing the word 'infrastructure' as a communicative device was repeatedly cited by those advocating a GI planning approach. Indeed, almost all those interviewed deemed use of the word 'infrastructure' as bequeathing discursive weight to green space issues in policy formulation by virtue of the word's association with other, more conventional interpretations of 'infrastructure'. However, the referential ambiguity of the term 'green infrastructure' enabled it to signify more than just the importance of ecological issues in green space planning. Rather, it could be employed as a linguistic mechanism to increase the weight of consideration given to numerous issues associated with green spaces. As noted by one planner,

> I think planners recognise that it's a brand, it's a concept which pulls together things that I suppose maybe planners have struggled in getting buy-in for, at an individual topic-by-topic level. Pedestrian networks, you know, really dull; 'Green Infrastructure' sounds much more interesting ... I think it may well be a mechanism by which to advance topics which traditionally individually might be quite difficult to do.
>
> (Interviewee C10)

This identification of GI as 'a brand' that addresses the problem of 'buy-in' for a variety of planning policy issues, suggests a degree of 'reflective practice' (Schön, 1991; Schön and Argyris, 1974) wherein new modes of representation were seen as necessary to effect change by both attracting attention to issues and legitimating perspectives regarding them (Gottweis, 2012; Hannigan, 2007; Laws and Rein, 2003; Duvall et al., 2017; Lennon, 2015). In particular, it was the perceived ability of the term 'GI' to cogently communicate the array of advocated benefits associated with green space planning that enabled the emergence and traction of a distinct GI narrative centred on the 'necessity' of such areas.

A 'Narrative of Necessity'

Those who advocated conceiving green spaces in terms of GI stressed the utility of the word 'infrastructure' in orientating audiences towards the vital services provided by such areas. In this sense, employing the word 'infrastructure' was seen as a way to present discussions on green space planning as a 'narrative of necessity'. It is against this backdrop that the following assertion was made,

> ... infrastructure is usually something you have to have and that's what I think is good about the term, is it sounds like something you need, you know, we need green infrastructure in this country sounds much better than saying we need green ways or we need biodiversity network [sic] ... it sounds more essential ... I mean it is essential but it's a term that sounds a bit more sort of business like and a bit more, you know, it's like your water infrastructure ...
>
> (Interviewee C3)

This ascription of necessity to green spaces by virtue of labelling them 'infrastructure' was viewed by those advocating GI as achieved by reconceptualising what was formerly perceived as 'the left over space' (Interviewee E4) as areas possessing essential society servicing functions. Thus, green spaces were reconceived as 'environmental resources' (Interviewee A2), which were seen to derive their value from the useful services or products they yield (Rees, 1990; Lennon and Scott, 2014; Potschin et al., 2016). Such a view was expressed by many of those interviewed, with one local planning authority official suggesting,

> ... we might describe schools and hospitals as social infrastructure and we have then kind of built, hard infrastructure like telecom systems or roads or water or sewers or whatever, so this green land then, you know, with this idea that it's not just sitting there doing nothing, is then another type of infrastructure.
>
> (Interviewee B20)

This new narrative of necessity, coupled with the familiar terminology of planning practice from which it arose, was thereby believed to lend greater weight of consideration to green space issues within discussions on appropriate landuse planning. Additionally, it was judged to facilitate the diffusion of this reconceptualisation both within and outside the institutions of planning governance. Such an opinion was reflected in the observation that,

> ... infrastructure is generally something that is required for an area. So by using the term green infrastructure it elevates it to be something that is required for an area. So it probably, I think it has taken off and I think it's, it's being more and more widely understood, within, certainly within planning and probably local government circles, and probably I think also in the community as well.
>
> (Interviewee B17)

Summary and Conclusions

This chapter argues that the initial impetus for introducing the term 'GI' into an Irish planning policy context stemmed from a desire to address perceived issues of ecosystem degradation resulting from habitat fragmentation. Also discussed was the widely held opinion regarding the difficulty in achieving this objective given the perceived low profile in planning policy formulation of ecological issues specifically, and green space issues more generally. Consequently, it is shown that those seeking to promote the consideration of ecological issues in planning policy formulation sought to establish a way to elevate the degree of consideration assigned to green space issues in landuse governance.

By virtue of widespread familiarity with the word 'infrastructure', and the connotations of indispensability ascribed to it, this chapter reveals how those advocating the allocation of greater emphasis to green space planning employed the word 'infrastructure'. Thus, 'GI' was employed as a linguistic device facilitating the reconceptualisation of green spaces from residual areas to locations providing crucial services to society. This enabled exponents of GI to fashion a 'narrative of necessity' with regard to such areas. To those advocating this approach, employing the term 'GI' was thereby viewed as a means to amass greater discursive weight for green space issues. Additionally, widespread familiarity with the word 'infrastructure', both within landuse governance institutions and as an element of common parlance, was viewed as prompting the reconceptualisation of green spaces as areas of 'use' to society.

As such, employing the term 'GI' as a means to advance various green space-associated issues reflects an appreciation that 'policy making is mostly a matter of persuasion' (Goodin et al., 2006, 5). Particularly, the opined 'discursive weight' garnered for green space issues by employing the GI 'brand' suggests that discourses concerning GI emerged in Ireland so as to cultivate the impression of green space as something socially, economically, environmentally and politically important (Hajer, 2003), as opposed to 'your left over, the stuff you haven't zoned'. In this context, the view that GI supplies a problem remedying 'proactive term' may be perceived as recognition that '[t]he struggle to define [a] situation, and thereby to determine the direction of public policy, is always both intellectual and political' (Schön, 1991, 348). Consequently, understanding the processes by which the meanings of GI were interpreted and advanced is central to an appreciation of how the GI policy approach emerged and evolved as a persuasive effort to address problematic policy ambiguity concerning green space associated planning issues. The next chapter therefore continues this interpretive analysis by focusing on the construal of GI's signification, significance and applicability. As such, it investigates how the 'meanings' of GI were constituted.

Notes

1 Comhar was the Irish Sustainable Development Council (SDC). It was dissolved in the winter of 2011. In January 2012, the sustainable development role formerly performed by Comhar was integrated into the work of the National Economic and Social Council (NESC).

2 Prior to its dissolution in 2011, Comhar, the Irish Sustainable Development Council, produced commentaries on a fortnightly or monthly basis. These provided a platform for those who were allied to Comhar to express their views on various aspects of sustainable development outside the formal confines of official documentation.

References

Barry, J. 2007. *Environment and Social Theory.* Oxford, England, U.K., Routledge.

Clabby, G. 2009. *Green Infrastructure: Critical Infrastructure for a Smart Economy* [Online]. Dublin, Ireland, Comhar SDC. [Accessed 12.11.10].

Comhar. 2010a. *Workshop on a Green Infrastructure Strategy for Ireland.* 8 February 2010. Dublin, Ireland.

Comhar. 2010b. *Workshop on The Economics of Ecosystems and Biodiversity.* 24 June 2010. Dublin, Ireland.

Comhar. 2010c. *Creating Green Infrastructure for Ireland: Enhancing Natural Capital for Human Well Being.* Dublin, Ireland, Comhar SDC.

DCC. 2005. *Dublin City Development Plan 2005–2011.* Dublin, Ireland, Dublin City Council.

Duvall, P., Lennon, M. and Scott, M. 2017. The 'natures' of planning: evolving conceptualizations of nature as expressed in urban planning theory and practice. *European Planning Studies*, 26, 480–501.

FCC. 2005a. *Fingal County Development Plan 2005–2011.* Dublin, Ireland, Fingal County Council.

FCC. 2005b. *Fingal County Heritage Plan 2005–2010.* Dublin, Ireland, Fingal County Council.

GCC. 2005. *Galway City Development Plan 2005–2011.* Galway, Ireland, Galway City Council.

Goodin, R. E., Rein, M. & Moran, M. 2006. The public and its policies. In: Goodin, R. E., Rein, M. and Moran, M. (eds.) *The Oxford Handbook of Public Policy.* Oxford, England, U.K., Oxford University Press.

Gottweis, H. 2012. Political rhetoric and stem cell policy in the United States: embodiments, scenographies, and emotions. In: Fischer, F. and Gottweis, H. (eds.) *The Argumentative Turn Revisited: Public Policy as Communicative Practice.* London, England, U.K., Duke University Press.

Hajer, M. 2003. A frame in the fields: policymaking and the reinvention of politics. In: Hajer, M. and Wagenaar, H. (eds.) *Deliverative Policy Analysis: Understanding Governance in the Network Society.* Cambridge, England, U.K., Cambridge University Press.

Hannigan, J. 2007. *Environmental Sociology.* New York, U.S.A., Routledge.

Laws, D. and Rein, M. 2003. Reframing practice. In: Hajer, M. and Wagenaar, H. (eds.) *Deliberative Policy Analysis: Understanding Governance in the Network Society.* Cambridge, England, U.K., Cambridge University Press.

Lennon, M. 2015. Explaining the currency of novel policy concepts: learning from green infrastructure planning. *Environment and Planning C: Government and Policy*, 33, 1039–1057.

Lennon, M. and Scott, M. 2014. Delivering ecosystems services via spatial planning: reviewing the possibilities and implications of a green infrastructure approach. *Town Planning Review*, 85, 563–587.

Potschin, M., Haines-Young, R., Fish, R. and Turner, R. K. 2016. *Routledge Handbook of Ecosystem Services*. Abingdon, England, U.K., Routledge.

Rees, J. 1990. *Natural Resources: Allocation, Economics and Policy*. London, England, U.K., Routledge.

Schön, D. A. 1991. *The Reflective Practitioner: How Professionals Think in Action*. London, England, U.K., Ashgate Publishing Limited.

Schön, D. A. and Argyris, C. 1974. *Theory in Practice: Increasing Professional Effectiveness*. San Francisco, California, U.S.A., Jossey-Bass Inc. Publishers.

Tubridy, M. and O Riain, G. 2002. *Preliminary Study of the Needs Associated with a National Ecological Network*. Wexford, Ireland, Environmental Protection Agency.

UCD, DLRCC, FCC and NATURA. 2008. *Green City Guidelines*. Dublin, Ireland, UCD Urban Institute.

Yearley, S. 2002. The social construction of environmental problems: a theoretical view and some not-very-Herculean labors. In: Dunlap, R. E., Buttel, F. H., Dickens, P. and Gijswijt, A. (eds.) *Sociological Theory and the Environment: Classical Foundations, Contemporary Insights*. Oxford, England, U.K.: Rowman & Littlefield Publishers Inc.

4   Naming Attributes and Effects

Entitlement

A 'correspondence theory' of truth takes 'scientific concepts to directly correspond to empirical referents of reality' (Fischer, 2003, 103). Here, '[s]cience is ideally a linguistic system in which true propositions are in one-to-one relation to facts' (Hesse, 1980, vii). However, this epistemological convention has been criticised for negating the 'constitutive' aspects of language, as it depicts 'language as a crucial instrument of knowledge, a very important representational tool, but nothing more' (Medina, 2005, 41). In contrast, a 'coherence theory' of knowledge emphasises the finite and temporally-bounded attributes of our comprehension of reality (Fischer, 2003, 103). Here, 'whether a belief is justified depends entirely on how well it fits or coheres with one's other beliefs, of its belonging to a coherent web of mutually supporting beliefs' (Lemos, 2007, 66). Such a perspective foregrounds a contextualised and linguistically-rooted comprehension of reality. Indeed, as early as the mid-twentieth century Peter Winch felt confident to proclaim, '[o]ur idea of what belongs to the realm of reality is given for us in the language that we use' (Winch, 1958, 15). This argument was subsequently advanced by Kenneth Burke when he suggested that the persuasively connotative functions of language usage may be considered as a process of 'entitlement'. Here Burke forwards,

> … a somewhat paradoxical proposition that experimentally reverses the common sense view of the relation between words and things. The common sense view favours the idea that 'words are the signs of things.' That is, various things in our way of living, are thought to be singled out by words which stand for them; and in this sense the words are said to be the 'signs' of those corresponding things. But if only as a tour de force, we here ask what might be discovered if we tried

DOI: 10.4324/9781003413608-5

inverting such a view, and upholding instead the proposition that 'things are the signs of words.'

(Burke, 1966, 360)

Burke continues his essay by laying emphasis on the selective and abstractive functions of naming by drawing attention to how the process of labelling simultaneously abbreviates the complex whilst also specifying the ontological status of something as, for example, an object, event, substance or vague feeling (Schiappa, 2003). A rhetorical effect of such 'entitlement' is that it creates the impression that what is entitled has always existed independent of its entitlement, and in a sense, was waiting to be discovered as the logical conclusion of investigations. As noted by Schiappa,

[n]aming and describing are acts of entitlement. Through such linguistic practices, we give our experiences meaning and make sense of reality. By entitling a given phenomenon, we locate that phenomenon in a set of beliefs about the world that includes beliefs about existence-status (what things are real or not) and essence-status (what qualities we may reliably predicate about the phenomenon). Because the range of possible entitlements is theoretically infinite, any given act of entitlement should be seen as a persuasive act that encourages language users to understand that-which-is-entitled in particular ways rather than others.

(Schiappa, 2003, 116)

This 'persuasive act' of entitlement is thus pertinent to an interpretive analysis of how representations of reality facilitate the emergence and evolution of a new policy approach. Of specific relevance to the present investigation of the ascension of policy in the absence of dispute is Schiappa's (2003) reasoning that in most cases the act of entitling proceeds unchallenged. Indeed, he asserts that it is only in hindsight that the persuasive function of a particular naming may become evident. In this instance, those advocating a particular entitlement are linguistically forming and communicating an interpretation of reality by offering a description that strategically functions in defining or redefining something without necessarily acknowledging that a new perspective is being promoted. Hence, such descriptions,

... are not claims supported by reasons and intended to justify adherence by critical listeners. Instead they are simply proclaimed as if they were indisputable facts.

(Zarefsky, 1998, 5)

In this sense, the term 'green infrastructure' (GI) may be understood as 'entitling' a form of reality that facilitates a reconceptualisation of green space, which in turn is perceived to bequeath it a higher profile in planning policy formulation. As is the case with the suggestion by Weaver (1985) that nouns in particular evoke pre-existence by suggesting a 'self-subsistent reality', this reality-recognising designation function (entitlement) of the expression 'GI' was alluded to by several interviewees. For example, one consultant planner engaged in the production of GI documentation noted,

> [i]n the same way that the concept of sustainability and sustainable principles and practices existed before someone said sustainability, green infrastructure did as well.
>
> (Interviewee A2)

Here, GI was understood to be applied retrospectively as a noun that designates what always existed. The term GI was thereby seen as a label that denoted an existing activity of landuse governance rather than representing a new concept or 'something we have created' (Interviewee B24). As such, the naming of 'GI' was understood to reflect an existing reality in harmony with a 'correspondence theory' of knowledge. However, closer scrutiny of how GI was interpreted suggests a 'coherence theory' of knowledge whereby readings of its meaning(s) were significantly influenced by how the semantic characteristics of the expression required its deciphering against a 'web of mutually supporting beliefs' (Lemos, 2007, 66). It was the perceived affinity or 'coherence' of such interpretations with existing ontological and epistemological commitments that induced the persuasive effects regarding the representation and constitution of the reality addressed by GI. Appreciating this phenomenon thus necessitates an investigation of the role played by 'naming attributes' in the meaning making process.

Naming Attributes

Many theorists argue that conceptions of reality are context dependent and rooted in the perspective of the interpreter (Bevir and Rhodes, 2015; Wagenaar, 2011; Rhodes, 2018; Ameel, 2020; Durnová, 2022; Fischer et al., 2015). As such, they contend that the 'facts' of policy rather than being objectively given are constructed and so may be more appropriately conceived as 'made'. In this sense, it is the construction of the reality through the manufacture of the facts to which policy is addressed that should interest the student of policy. Comprehending the construction of such policy facts through a linguistically-'entitled' reality requires careful consideration of the communicative requirements of the entitlement process.

With regard to the term 'GI', a number of those interviewed felt that although clearly denoting something that currently exists, and connoting an idea of something 'necessary', it did not immediately refer to an obviously defined entity. Instead, interviewees suggested that what the term signified may initially seem ambiguous to the interpreter. This attribute of *ambiguous signification* was summarised by one QUANGO official who concluded,

> [t]here's different slants on it … you could look at green infrastructure … things like wind veins and wind farms and I suppose facilities that would relate to energy, would relate to biodiversity, would relate to heritage, culture etc, etc, there's a variety of probably spins you could put on it. So that's why I suppose … it's a complex term and it can be, it's a bit ambiguous.
>
> (Interviewee C4)

In the case of GI, this attribute of *ambiguous signification* meant that reaching apparent clarity of interpretation necessitated reasoning what the expression represented by exploring its connotations. This interpretive exercise was explained by Roland Barthes (2009[1957]) in terms of levels of meaning[1] whereby that indicated by the sign (e.g. 'infrastructure') itself becomes the signifier for something else by way of association (e.g. something 'necessary'). Although such *associative interpretation* 'works on the subjective level' (Fiske, 1990, 87), Chandler notes that,

> [i]ntersubjective responses are shared to some degree by members of a culture; with any individual example only a limited range of connotations would make any sense. Connotations are not purely personal meanings – they are determined by the codes to which the interpreter has access.
>
> (Chandler, 2002, 139)

Fiske outlines how such intersubjective responses to interpretation mean that 'it is often easy to read connotative values as denotative facts' (1990, 87). It was this feature of associative interpretation which led Barthes (1990[1974]) to conclude that connotation may induce the illusion of denotation. In this context, the transition from connotation to apparent denotation is conceived as a process of 'naturalisation'. Here, the powerful impression of literal denotation masks the connotative readings intrinsic to the sign's comprehension (Chandler, 2002). Such construal of meaning is facilitated by the 'contextual determinacy' of interpretation (Gadamer, 2004; Medina, 2007; Stern, 2008; Wittgenstein, 1953), wherein words uttered in a particular context do not elicit a range of connotations. Rather, such words call forth only the 'contextual connotations' of the words used (Beardsley, 1981 [1958], 125).

In the case of GI, the intersubjective connotative reading of green 'infrastructure' as something that 'isn't just a potential discretionary or stylistic approach' (Interviewee A7), but rather, as 'something you have to have' (Interviewee C3), prompted a sense of necessity in the associative interpretation of an otherwise ambiguous term. Indeed, the potency of such connotations related to the word 'infrastructure', and the common familiarity with such connotations, elicited a sense of literal denotation of the expression 'GI' that partially concealed the processes of interpretation required by its entitlement. This deduction of meaning from an ambiguous expression via such 'associative interpretation' was alluded to in several interviews and lucidly conveyed in the opinion,

> I think it's a very practical word and it conveys the idea of the services I think very well because we are able to make that direct link between like our waste water systems and all this as being part of our infrastructure. Even things like our hospitals, our schools, all those things, that these are things that we need. We can't live without them. We can't live the life we currently live without these things and they don't just occur by accident, we have to plan them. We have to know what our populations are going to be, we have to know who's going to be living where … you have to organise them, you have to plan for them. So I think the 'infrastructure' side of it helps to convey that and it helps in the explaining of the term.
>
> (Interviewee B1)

This opinion that the word 'infrastructure' helps explain the meaning of the term GI presumed the likelihood of shared interpretations of GI's signification. Such a supposition blurred the boundaries between connotation and denotation. In doing so it echoed concepts theorised by Berger and Luckmann on the intersubjective projection of interpretations in the generation of an objective reality. Here, 'the fundamental legitimating "explanations" are … built into vocabulary' (Berger and Luckmann, 1966, 112). However, this social construction of reality (entitlement) was not neutral. Indeed, the interpretive requirements of GI entailed mediating meaning through context contingent linguistic conventions (Lemke, 1998). In this sense, the connotations that helped to convey the meaning of GI may carry with them associations beyond those of 'necessity'. This is consequent on comprehending green spaces through the prism of more familiar 'infrastructure' wherein planning activities are conducted against a backdrop of particular ontological and epistemological presuppositions regarding reality. In other words, naming has effects.

Naming Effects

The Role of Metaphor

The word 'green', as used in the expression 'GI', is identifiable as a specific type of metaphor termed 'metonymy'.[2] As a non-literal expression, metonymy operates via 'the evocation of the whole by connection' (Chandler, 2007, 130). While normal metaphors are literally impossible, 'the grounding of metonymic concepts is, in general, more obvious than is the case with metaphoric concepts' (Lakoff and Johnson, 2003, 39), since there is 'some observable, often physical, connection between metonymy and its meaning, whereas metaphors rely on comparisons of sorts' (Knowles and Moon, 2006, 9). This difference can lead metonymy to 'seem more natural than metaphors' (Chandler, 2007, 132). As such, metonymic reasoning insinuates the 'grounding' (Lakoff and Johnson, 2003, 39) of a concept by way of its connection to experience, unlike the 'imaginative leap' (Chandler, 2007, 132) required by normal metaphor. Metonymic reasoning thus adheres to Barthes' hypothesis on the 'naturalisation' of connotation by enabling the referenced term to appear denotative despite its capacity to accommodate multiple connoted meanings. In the case of the term 'GI', the word 'green', with its popular use as a prefix and suffix for political, economic and social activities perceived as promoting environmental sensitivity (Carter, 2007; O'Neill et al., 2008; Hansen and Cox, 2015), not only metonymically connotes activities that specifically address environmental protection, but also the spaces normally labelled 'green'. The former interpretation is illustrated by assertions such as,

> I suppose green connotates [sic] living environment, maybe clean to people, maybe sustainability, maybe low energy or those kind of connotations come with green.
>
> (Interviewee B5)

Nevertheless, in the context of GI, it was as a reference to green spaces that most of those interviewed interpreted 'green' as signifying. However, as noted by many interviewees, the scope of spaces represented by the use of 'green' in the context of the expression 'GI' was abundant. This was coherently expressed by one planning authority officer, who suggested,

> ... the word green, it can encompass anything to do with the natural environment ... So when you're talking about green you could be talking about golf courses, you could be talking about park lands, you could be talking about the open countryside, you know. It gives you broad scope I suppose to examine the area that you want to.
>
> (Interviewee B2)

Thus, whereas most interviewees consider 'green' as signifying a type of 'space', the metonymic qualities of the word acted as 'a primary source of polysemy' (Gentner and Bowdle, 2008, 119), wherein the criteria for topographic relevance were unspecified. The forms of space signified by the word 'green' were not defined. Instead, they were consequent on subjective interpretation. Latitude for interpretation of the word 'green' thereby provided scope for the application of GI to varied spatial typologies within landuse planning.

External to the expression 'GI', the word 'infrastructure' was a noun seen to designate,

> ... the building blocks for planning and for designing towns and framing investment and so you have transport infrastructure, water services infrastructure ...
>
> (Interviewee B16)

Although what the word 'infrastructure' signifies was not circumscribed, there existed broad consensus among those interviewed that it connotes something that directly facilitates society's economic and physical maintenance and growth – something 'necessary'. Thus, assembling the words 'green' and 'infrastructure',

> ... bends the understanding a little bit, but I suppose that's where you get the green, the two, the green and the infrastructure coming together. That sort of grabs people alright and you know, it's possible to, to build it into, to the context of sort of grey infrastructure, IT infrastructure and so on. All of which are very sort of concrete, sort of visible things on the ground.
>
> (Interviewee A4)

The conjunction of these words generated an interpretive 'bridge that allows passage from one world to another' (Shiff, 1978, 106), in which 'the reference of the metaphorical statement [has] the power to "rede-scribe" reality' (Ricoeur, 2002[1975], 5). Understanding how this metaphor fostered a reconceptualisation of green space necessitates an appreciation of the way the two words of 'green' and 'infrastructure' were asymmetrically positioned relative to each other in terms of how they perform their meaning-endowing functions. To achieve this, a consideration of the mechanics of interpretation is required.

Organising Interpretation

Ivor Richards (1965[1936]) proposed the comprehension of metaphor as the unity of an underlying idea with the means employed in its conveyance. The former he terms the 'tenor', while the latter he refers to as the 'vehicle'. As discussed in the previous chapter, the idea that the advocates of GI sought to convey was the importance of green spaces in sustaining and facilitating society-centred development while con-currently enabling environmental conservation. The vehicle used to com-municate this tenor (idea) was the expression 'GI'. It is in this sense that several interviewees considered the benefit of utilising the term GI as,

> [i]t puts a name, a label on something, a concept that we might be trying to achieve ... using the term green infrastructure might actually put some sort of term on it that people who aren't neces-sarily of that natural way of thinking could actually start to ima-gine it, or visualise it, and see possible benefits and services and values to that.
>
> (Interviewee C7)

However, as stressed by Paul Ricoeur, '[t]he metaphor is not the vehicle alone but the whole made of the two halves' (Ricoeur, 2002[1975], 93). In this context he explicates how,

> [t]he simultaneous presence of the tenor and vehicle and their interaction engender the metaphor; consequently, the tenor does not remain unaltered, as if the vehicle were nothing but wrapping and decoration.
>
> (Ricoeur, 2002[1975], 93)

As a form of complex metaphorical entitlement, it follows that use of the term 'GI' (vehicle) to convey the importance of green space (tenor) not only achieved the manifest objective of the communicative act, but also altered perceptions on how the significance of green space was conceived. Black (1962) suggests that this alteration transpires through the work of metaphor in 'organising' our interpretation of what is being conveyed (tenor) through means of emphasis and suppression. This 'interaction view' of metaphor (Hausman, 2006, 229) describes how a metaphorical word or expression 'gains new meaning' (Lyon, 2000, 138). Ricoeur concisely explains such a phenomenon by noting that this is achieved via metaphor '[o]rganising a principal subject by applying a sub-sidiary subject to it' (Ricoeur, 2002[1975], 101). With regards to GI, the principal subject 'organised' was the word 'green', while the subsidiary subject engaged in organising was the word 'infrastructure'.[3] The effect of this organising of interpretation was coherently outlined by one planner who commented that,

> [i]nfrastructure is like an underlying framework for a particular system or feature of a system. So basically what you're looking at is the idea of green in terms of, well green areas, green spaces or whatever you want to encompass in the term green and then putting that in a context so you actually have a framework for developing or understanding a methodology or an approach to developing the idea of how you use these spaces or areas and what you use them for. So when you put the two of them together you know, you actually do get quite a useful phrase in terms of creating infrastructure ...
>
> (Interviewee B2)

Here a 'system of associated implications' (Black, 1962, 39–40) was transferred from the familiar understandings of 'infrastructure' onto interpretations of 'green' as a spatial referent. Hence, an 'emergent meaning' (Beardsley, 1981[1958], 131) of 'green spaces' was prompted wherein such areas were seen to serve a development-linked purpose that should be planned in accordance with the methods normally associated with conventionally-conceived infrastructure. Thus, forging the metaphor 'GI' enabled the configuration of specific ontological, epistemological and functional interpretations as to the nature of green spaces ('green'). In this sense, 'GI' became a *conceptual metaphor*.

In their seminal study of metaphor's capacity to direct thought, Lakoff and Johnson (2003) identify three categories of conceptual metaphors: ontological, structural and orientational. Whereas orienta-tional metaphors (up, down, left, right, etc.) do not concern the current study, an awareness of ontological and structural metaphors is crucial to an understanding of GI's role in the reconceptualisation of green spaces. Ontological metaphors enable the conceptualisation of 'things, experiences and processes, however vague and abstract, as if they have definite physical properties' (Knowles and Moon, 2006, 40). Structural metaphors facilitate the structuring of one concept in terms of another. Conceptual metaphor theorists hypothesise that metaphors form systematic sets of correspondences, or 'mappings' across conceptual domains (Semino, 2008), where the 'source domain' is used to describe the concept area from which the metaphor is drawn, and the 'target domain' is used to identify the concept area to which the metaphor is applied (Knowles and Moon, 2006). Under this model, source domains supply frameworks for target domains, which subsequently determine the manner by which the entities of the target domains are conceived and discussed (Lakoff, 1993; Schön, 1993).

Applied to the term 'GI', such an understanding would suggest the organisation of the principal subject of green spaces ('green') by means of the subsidiary subject ('infrastructure'), via mapping associations from commonly-conceived notions of infrastructure (source domain) onto comprehensions of green space ontology and an associated epistemology of green space planning (target domain). Thus, as conveyed by one local authority planner,

> [w]hat the two words 'green infrastructure' are, taking the second word first, I suppose that suggests that you look upon these areas as areas that are, they're part of the infrastructure so in the same way that roads are infrastructure, community services are infrastructure, then green infrastructure is the amenity and the green in recreational areas where our local communities can use to enjoy, and they contribute to the quality of life of a town or a city.
>
> (Interviewee B17)

Mapping associations of conventionally-conceived 'infrastructure' onto 'GI' involved a process of *patterning* in the concept transference from the source to target domains. Several forms of metaphorical patterning have been theorised (Goatly, 1997; Kövecses, 2002; Knowles and Moon, 2006; Lakoff and Johnson, 2003; Semino and Demjén, 2016). Many of these entail overlap and co-occurrence, and as such can be reduced to an interrelated co-operative pair. These are namely *repetition* and *recurrence*.

'Repetition' involves little more than the reiteration of particular expressions in discourse. Nevertheless, repetition can be a powerful device in establishing an 'evaluative accent' (Maybin, 2001, 65) on that which is metaphorically conceived. This is illustrated, for example, by the repeated use of the noun 'benefit' and the phrase 'quality of life' with regards to GI. Both the documentary and interview data indicate frequent instances of this form of repetition, such as expressed by the assertion that,

> [i]t's looking at natural places, like not just for wildlife conservation but for the *benefits* that it gives people in terms of you know, just a better feeling, quality of life and obviously health *benefits* and yeah, amenity and recreational *benefits* as well.
>
> (Interviewee B19) [Emphasis added]

Separate to repetition, but often working in parallel, 'recurrence' entails the use of different expressions relating to the same broad source domain when expanding upon a discussion and/or conceptualisation of a target domain. Thus, recurrence is an important aspect of metaphorical analysis as it is indicative of how some meaning components of the source domain are highlighted in constituting the ontologies and epistemological attributes of the target domain. In the case of GI, the particular forms of recurrence were both influenced and facilitated by the pre-existence of several 'networked' approaches to planning predating the re-emergence of the GI discourse in 2008. Thus, for example, by employing terms such as 'network', 'link' and 'connectivity', the GI discourse drew upon the lexicon of existing 'networked' green space planning discourses and amalgamated these with the diction of familiarly perceived 'infrastructure'. Cases of such recurrence were regularly expressed in interviews, as illustrated in the opinion that,

> [i]t's [GI] about the *connectivity* of green spaces and the whole range of green spaces as they are in terms of coastal strips, in terms of river corridors, in terms of *networks* of parks and spaces and even issues like green roofs and how you *link those together*. So I think it's about *connectivity* and creating that *connectivity* you know what I mean.
>
> (Interviewee E4) [Emphasis added]

Words such as 'management', 'engineering' and 'services' expanded this lexicon by transferring terms from the source domain of traditionally-conceived infrastructure into the target domain of reconceptualised green space purposes and planning. This construed green space, and the issues associated with it, as primarily concerned with facilitating the provision of society-centred services. Instances of such recurrence were well represented in GI documentation, for example,

[t]he Green Infrastructure concept involves the *planning, management and engineering* of green spaces and ecosystems in order to provide specific *benefits to society.*

(UF and IEEM, 2010, 2) [Emphasis added]

Therefore, by employing the 'GI' metaphor to communicate the advocated importance of green spaces, those promoting this discourse provoked the reconceptualisation of these areas in terms of conventionally conceived (networked) infrastructure. Consequently, this altered perception of green space modified reasoning on its function, and as a corollary transformed opinions on the purposes and appropriateness of planning approaches to those issues associated with such areas.

Functional Expectations

The reconceptualisation of green spaces stimulated by the term 'GI' appeared to establish expectations regarding the functions such areas were seen as appropriately delivering. Thus, in contrast to perceiving such locations as 'the left over space … the stuff you haven't zoned' (Interviewee E4),

> … from a planning perspective if you're looking at the environment and you have a map which would usually just be a habitat map, instead of just saying what it is, it's also what it does. So it's not just the woodland, a river corridor; it's about carbon sequestration, it provides fuel, it provides flood amelioration, it provides water, all of these things …
>
> (Interviewee C8)

In this sense, both the expression 'GI' and the advocacy discourse in which it was employed, prompted a principal concern with green space uses to society, while concurrently reducing the attention accorded to the potential significance of such areas for functions other than those serving human needs (Lennon, 2015). While some concern was still evident for non-human interests, the perceived value of green spaces became primarily anthropocentric and instrumental. This orientation was frequently articulated by both Irish non-statutory planning publications advocating a GI planning approach (Comhar, 2010; FI, 2010; HC, 2010; UF and IEEM, 2010), and those which statutorily promoted GI by way of planning policy provisions (DCC, 2010; DLRCC, 2011; SDCC, 2010). Thus, an immanent critique would suggest that the ontological, epistemological and functional interpretations of green spaces were reorientated to benefit the maintenance of the built environment and facilitate economic growth. This focus on selective anthropocentric utility eclipsed an ecocentric perspective on the conservation of such areas that provided the initial impetus for introducing the term GI.

An Interpretive Triad

Connotative Reasoning

The widespread familiarity of the word 'infrastructure' and its normative inferences engendered a construal of GI as that which 'should be viewed as critical infrastructure for Ireland in the same way as our transport and energy networks are as vital to sustainable development' (Comhar, 2009, 39). Accordingly, those advocating GI envisaged it as a 'strategically planned and delivered network of high quality green spaces and other environmental features ... designed and managed as a multifunctional resource' (SDCC, 2010, 257). In this context, the activity of GI planning was perceived through the prism of conventionally-conceived infrastructural planning whereby '[t]he Green Infrastructure concept involves the planning, management and engineering of green spaces and ecosystems in order to provide specific benefits to society' (UF and IEEM, 2010, 2).

Resultant from such connotative reasoning was the presumption that the 'Green Infrastructure approach to planning is grounded in sound science, spatial landuse planning theory and practice' (Comhar, 2010, 59). Consequently, GI planning was seen to entail the deployment of 'the old processes of survey, analysis, plan' (Interviewee B17) as the methodologies normally associated with the assessment and design of conventional infrastructure. Central to this logic was the role played by mapping in giving denotative potency to the connotatively reasoned act of entitlement. This helped to instigate what Barthes (1990[1974]) labels *naturalisation* in the perception of GI as a concept that 'kind of ties back into common sense in a way' (Interviewee B13), so that when introduced 'it seemed to make sense to planners and landscape architects and spatially minded people' (Interviewee B20). This led many of those interviewed to remark that GI 'just makes sense' (Interviewee B16), and as outlined by one consultant, to conclude,

[w]hen I first came across the term ... it was far from something alien, it was in fact something very familiar. It was really very, almost totally familiar to me so I took great heart from the, from

learning that this was a, more than just a term but actually it represented a school of thought.

(Interviewee A7)

Such perceptions of familiarity reflect the assertion of Berger and Luckmann that,

[w]hat is real 'outside' corresponds to what is real 'within'. Objective reality can readily be 'translated' into subjective reality, and vice versa. Language of course, is the principal vehicle of this ongoing translating process in both directions.

(Berger and Luckmann 1966, 153)

In this sense, the *naturalisation* of GI through perceptions of it as 'something very familiar' and of 'common sense' was effected through agent projection of assumptions regarding what constitutes 'proper planning process' (Interviewee A10), and as a corollary, proper professional practice. What was perceived to constitute legitimate, and thereby 'proper planning process', was delineated by the modernist rationalities of landuse governance (Fischer, 2003; Flyvbjerg, 1998; Rydin, 2003). Here, legitimate practice 'is highly profiled in the sense of representing fully the objective reality within which it is located' (Berger and Luckman, 1966, 184). Thus, 'one rhetorical effect of entitling a new "thing" is that it creates the impression that the thing [GI] has been "out there" all along' (Schiappa, 2003, 115). It is in this sense that the particular forms of *connotative reasoning* involved in the entitlement of 'GI' prompted assumptions of it as familiar planning practice that 'just makes sense' (Interviewee B16). Specifically, such *connotative reasoning* induced perceptions of GI as possessing characteristics akin to that of conventionally-conceived 'infrastructure'. GI planning was thereby constituted as a rationally-justified policy approach that emphasised the anthropocentric utility of green spaces, and by association, nature.

Apparent Simplicity

Intrinsically related to the concept of naturalisation resulting from connotative reasoning were impressions of *apparent simplicity* in understanding what GI meant. As noted by one consultant,

I think most people that hear about it, it sort of clicks a light on in their head and they go ah yeah, it's kind of self-evident; that seems like a good idea ... So I'm guessing that's why it's beginning to take a foothold. People are going, 'ah yeah, that seems to make sense, let's try and do that'.

(Interviewee A2)

Likewise, the logic employed in GI planning was perceived to be easily comprehensible given that it was viewed to equate with the methods employed in the planning of traditionally conceived 'infrastructure'. As noted by one planning authority officer,

> I think this is just good planning practice: so it's like map what you have, so find out what you have … then think about what you need into the future and then see how you go about managing what you have or providing new stuff to fulfil that need. So that's kind of the ideal scenario is that you would see what you have, look at its functions and benefits it provides, see what functions and benefits you need into the future and then either upgrade what you have or manage better what you have and then provide new stuff.

(Interviewee B20)

The evocations inherent to connotative reasoning thereby shaped epistemological perspectives regarding the *apparent simplicity* of the logic adopted in response to the entitled reality. This phenomenon was discussed by Boyd (1993) regarding what he termed the 'exegetical' and 'theory constitutive' potentials of metaphor. Exegetical metaphors are important in the pedagogical conveyance and dissemination of an idea. Their essential characteristic is that they are dispensable, since the theorists employing them have non-metaphorical means of expressing and referring to the same phenomenon (Semino, 2008). In contrast, 'theory constitutive' metaphors are defined by their function in 'the development and articulation of theories in relatively mature sciences' (Boyd, 1993, 482). Such metaphors provide a vocabulary in which to perceive new concepts within the existing discursive field of established disciplines (Haack, 1987–1988; Hausman, 2006). Boyd's distinction is thus best viewed not as capturing two dissimilar categories of metaphor, but rather two different functions that metaphors can perform when employed for specific functions at particular points in the historical development of an idea (Semino, 2008, 134). Accordingly, although it is possible for a metaphor to be used exclusively for either exegetical or theory-constitutive purposes, Semino notes,

> … it is often the case that the 'same' metaphor may have a primarily theory-constitutive function in one context and a primarily

educational function in another, or may perform both at the same time.

(Semino, 2008, 134)

Regarding the connotative reasoning necessitated in the interpretation of 'GI', the interplay between 'theory constitutive' and 'exegetical' influences in the entitlement of the concept was represented by views on the *apparent simplicity* of its understanding comparable to that of conventionally-conceived 'infrastructure'. This formed a reciprocal arrangement wherein the *apparent simplicity* of the concept was facilitated by its connotatively-reasoned constitution, which in turn symbiotically assisted its explication. In this way, the boundaries between connotation and denotation became blurred as the GI metaphor concurrently served as both a model 'of' a situation and a model 'for' it (Yanow, 2000, 43). As concluded by one planner,

> [t]here's no difficulty in understanding the concept when you explain to people; well it's the same as the way we plan for development or any kind of development, it's just being a bit more proactive as to how we develop our green areas and how we care for natural pieces of infrastructure, green infrastructure that are there, and how we create new ones as well.
>
> (Interviewee C5)

Such impressions of *apparent simplicity* served to supply an apparent clarity of meaning and direction for the resolution of problematic policy ambiguity. This resulted from how 'the use of theory-constitutive metaphors represents a nondefinitional reference-fixing strategy' (Boyd, 1993, 496). In doing so, the *apparent simplicity* engendered by the GI metaphor presented a heuristic tool (Black, 1962, 84) that invited the interpreter 'to explore similarities and analogies between features' (Boyd, 1993, 489) while simultaneously not delineating that which was discussed. Latitude for subjective interpretation was thereby facilitated through the 'emergent meaning' (Beardsley, 1981[1958], 131) of GI wherein an 'inductive open-endedness' was permitted (Boyd, 1993, 488). The impression that '[t]here's no difficulty in understanding the concept when you explain [it] to people' was thus encouraged by the capacity of interpreters to maintain unchallenged perceptions of accuracy regarding their (potentially divergent) ideas concerning what GI may signify. Consequently, the processes of connotative reasoning that evoked perceptions of *apparent simplicity* in GI's signification as a form of 'infrastructure' paradoxically functioned as a 'semantic mechanism for creating and extending polysemy' (Medina, 2005, 127), wherein the format and beneficiary of such 'infrastructure' was left unspecified. In this way, the *apparent simplicity* in comprehending GI existed in a mutually-dependant relationship with the term's *flexible signification.*

Flexible Signification

The attribute of ambiguous signification inherent to the *entitlement* of GI also provoked the phenomenon of *flexible signification* in the term's application. Resulting on the requirement to interpret the expression via connotations with commonly conceived 'infrastructure', there existed a degree of polysemantic latitude in the meanings attributed to 'GI'. Although GI induced perceptions of *naturalisation* in denoting an idea conceptually tethered to traditionally understood 'infrastructure', the interpretive requirements of connotative reasoning necessitated the investment of subjective appreciations of what such commonly-conceived infrastructure entailed. In this sense, GI simultaneously encompassed a multiplicity of signified ideas and objects normally distinguished as distinct entities. As noted by Schaffer,

> [t]he various uses or meanings of a word do not interlock precisely like pieces of a jigsaw. Consequently, to say that we can identify shared meanings implicit in a word is not to claim that those meaning can be arranged tidily. A word can be used in a variety of different, and sometimes contradictory, ways (even by one person, in one conversation).
>
> (Schaffer, 2006, 153)

This potential to concurrently encompass a variety of infrastructure-associated ideas and objects was frequently expressed in interviews, leading one planner to conclude that 'it's a bit like the big bang you know, the longer it goes on the more diverse it gets in its meaning and application' (Interviewee C5). Similarly, another planner involved in the production of GI planning guidance observed that,

> [i]t [GI] includes all these kind of things: biodiversity management and enhancement, water management, drainage, flood attenuation, filtration, pollution control, recreation, tourism, visual amenities, sense of place, sustainable mobility, food, timber, other primary products, regulation of microclimates.
>
> (Interviewee B12)

Likewise, a different interviewee surmised,

> I probably would take the view that green infrastructure is nearly everywhere in a way. As I say you can take particular things be it a disused railway track we are now converting to a walkway or whatever or a cycleway or a river bank, canal tow path, harbour, a beach, lake. Even the motorway, I kind of tend to be all encompassing because infrastructure is everywhere.
>
> (Interviewee C9)

Such flexibility of signification required the imposition of 'judgement' (Ricoeur, 2002[1975], 66) in the interpretation of GI's connotative potential. This mediation of meaning by connotative reasoning could not have been objective, as it obliged the interpreter to subjectively invest that which was being interpreted with a signification it did not already possess by way of existing formal denotation. In this context, 'policy analysts are situated knowers thinking and writing from particular points of view' (Yanow, 1996, 27). This capacity for the term to be 'positioned' (Hajer, 2003) relative to the perspective of the interpreter led several of those interviewed to deduce 'GI',

> ... as a generic catch all. Sometimes it's trying to address water supply issues. Sometimes it's trying to address energy issues, sometimes transport issues, so it depends on the context, depends on the person. It depends on the function.
>
> (Interviewee E5)

Such *flexible signification* facilitated appropriation of GI's entitlement for the particular needs of the end user. As noted by one local authority planner,

> I think the key thing for anybody to realise is there's no definition of green infrastructure and I think that's so important ... it's whatever the hell you need in your area.
>
> (Interviewee B24)

Thus, the expression 'GI' defined not an entity or idea tightly delineated in possible application, but rather something loosely circumscribed by connotations with traditionally-conceived infrastructure, whose quality of *flexible signification* enabled latitude in its use. The purposes to which it was put were therefore as much dependent on the objectives of those using it as they were on the meanings it was seen to imply. Such *flexible signification* thereby operates in a relationship of reciprocity with *connotative reasoning* and *apparent simplicity* as a triad of 'naming effects' prompted by the entitlement of GI and giving meaning to its expression. This relationship is illustrated in Figure 4.1.

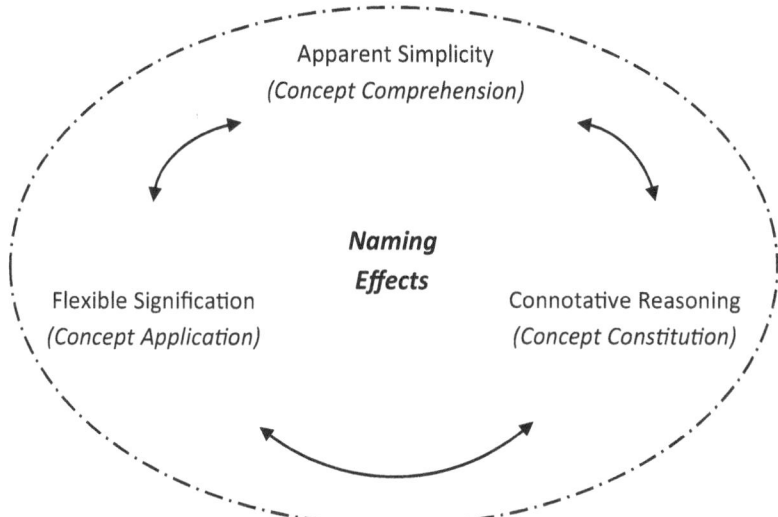

Figure 4.1 Diagrammatic Representation of the Relationships between the 'Naming Effects' of GI's *Entitlement*
Source: author

Summary and Conclusions

This chapter builds upon the analysis of the previous chapter which examined the reasons why the term GI was introduced and how this was employed to produce a 'narrative of necessity'. The present chapter extends this investigation by studying how such a narrative was generated through interpretations of the term 'GI'. The chapter provided an examination of how the entitlement of GI stimulated particular ontological and epistemological assumptions. This interpretation was prompted by the ambiguous signification of the term and a requirement to metaphorically interpret its meaning by way of association with familiar ideas and objects. Such requirements prompted a number of 'naming effects' in the constitution of the concept, its explication, and its application. Rather than operating in isolation, these effects formed a triad of mutually-dependent and reinforcing processes that gave meaning to the expression 'GI'.

Demonstrated is that these meaning making processes served more than simply constituting 'GI' as a reflection of an objective reality. Rather, such interpretive processes had an ontological function in which 'the reference of the metaphorical statement [had] the power to "redescribe" reality' (Ricoeur, 2002[1975], 5). Hence, the seemingly objective world related by GI was actually subjectively and inter-subjectively constituted 'through' readings of GI's meaning. In this sense, the 'emergent meaning' (Beardsley, 1981[1958], 131) prompted by GI's interpretation may be conceived as producing a new ontology of green spaces, and as a corollary, nature (Castree, 2014; Duvall et al., 2017; Kohout and Kopp, 2020; Matsler et al., 2021).

However, it is shown that the 'emerging ontology' of such spaces was not neutral (Hacking, 2002). As noted by Yee (1996, 97), 'meanings quasi-causally affect certain actions not by directly or inevitably deter-mining them but rather by rendering these actions plausible or implausible … respectable or disrespectable'. Thus, understandings of both the possibility and appropriateness of planning activities regard-ing green spaces, and by association 'nature', became established rela-tive to how this world was conceived in the formulation of landuse policy. In terms of GI's ascension as a policy approach, the emerging ontology of green space was manifested in the equation of such areas, and nature more generally, with conventionally conceived 'infra-structure'. As a result of this reconception, the objectives and activities of landuse governance were directed away from the view that the environment 'must be protected or conserved at all cost' (Interviewee A2). In its place emerged a belief that proper green space planning entailed the design and delivery of environmental services to facilitate society's requirements. Consequently, the perceptions of 'use in this green stuff' (Interviewee A5) provoked by GI during this period were for the most part anthropocentric and concerned instrumental value. Appreciating how such understandings achieved purchase among the landuse planning fraternity necessitates an investigation of how inter-pretations of GI's meaning resonated with the rationalities of planning practice. Thus, the subsequent chapter examines the role of 'rationality resonance' in the emergence and evolution of the GI planning policy approach in Ireland.

Notes

1 In discussing the signification characteristics embodied in particular narra-tive forms (myths), Barthes refers to a 'second order semiological system' of connotation (Barthes, 2009[1957], 37–38).

2 As there is some debate regarding the subdivision of 'metonymy' into a dyad of 'metonymy' and 'synecdoche', the term 'metonymy' will be employed here as encompassing both terms. Although Chandler defines synecdoche as, 'a figure of speech involving substitution of part for whole, genus for species or vice versa' (Chandler, 2007, 262), he notes, '[e]ven if synecdoche is given a separate status, general usage would suggest that metonymy would remain an umbrella term for indexical links as well as having a narrower meaning of its own' (Chandler, 2007, 132–134).

3 It is acknowledged that scope exists for this process to work in reverse. However, the empirical research indicated that the organisation of the word 'green' by the word 'infrastructure' predominated with respect to GI in Ireland.

References

Ameel, L. 2020. *The Narrative Turn in Urban Planning: Plotting the Helsinki Waterfront*. Abingdon, England, U.K., Routledge.

Barthes, R. 2009[1957]. *Mythologies*. London, England, U.K., Vintage.

Barthes, R. 1990[1974]. *S/Z*. Translated 1974. Oxford, England, U.K., Basil Blackwell Ltd.

Beardsley, M. C. 1981[1958]. *Aesthetics: Problems in the Philosophy of Criticism*. Indianapolis, Indiana, U.S.A., Hacket Publishing Company.

Berger, P. and Luckmann, T. 1966. *The Social Construction of Reality: A Treatise on the Sociology of Knowledge*. London, England, U.K., Penguin Books.

Bevir, M. and Rhodes, R. A. W. 2015. *Routledge Handbook of Interpretive Political Science*. Abingdon, England, U.K., Routledge.

Black, M. 1962. *Models and Metaphors*. Ithaca, New York, U.S.A., Cornell University Press.

Boyd, R. 1993. Metaphor and theory change: what is 'metaphor' a metaphor for? In: Ortony, A. (ed.) *Metaphor and Thought*. Cambridge, England, U.K., Cambridge University Press.

Burke, K. 1966. *Language as Symbolic Action*. Berkley, California, U.S.A., University of California Press.

Carter, N. 2007. *The Politics of the Environment*. Cambridge, England, U.K., Cambridge University Press.

Castree, N. 2014. *Making Sense of Nature*. London, England, U.K., Routledge.

Chandler, D. 2007. *Semiotics: The Basics*. London, England, U.K., Routledge.

Comhar. 2009. *Towards a Green New Deal*. Dublin, Ireland, Comhar SDC.

Comhar. 2010. *Creating Green Infrastructure for Ireland: Enhancing Natural Capital for Human Well Being*. Dublin, Ireland, Comhar SDC.

DCC. 2010. *Dublin City Development Plan 2011–2017*. Dublin, Ireland, Dublin City Council.

DLRCC. 2011. *Proposed Variation No. 2 to the Dún Laoghaire Rathdown County Development Plan 2010–2016 (Sandyford Urban Framework Plan)*. Dublin, Ireland, Dún Laoghaire Rathdown County Council.

Durnová, A. 2022. Making interpretive policy analysis critical and societally relevant: emotions, ethnography and language. *Policy & Politics*, 50, 43–58.

Duvall, P., Lennon, M. and Scott, M. 2017. The 'natures' of planning: evolving conceptualizations of nature as expressed in urban planning theory and practice. *European Planning Studies*, 26, 480–501.

FI. 2010. *Historic Towns in Ireland: Maximising Your Tourist Potential.* Dublin, Ireland, Fáilte Ireland.

Fischer, F. 2003. *Reframing Public Policy: Discursive Politics and Deliberative Practices.* Oxford, England, U.K., Oxford University Press.

Fischer, F., Torgerson, D., Durnová, A. and Orsini, M. 2015. *Handbook of Critical Policy Studies.* Cheltenham, England, U.K., Edward Elgar Publishing Limited.

Flyvbjerg, B. 1998. *Rationality and Power: Democracy in Practice.* London, England, U.K., The University of Chicago Press Ltd.

Gadamer, H.-G. 2004. *Truth and Method.* London, England, U.K., Continuum International Publishing Group.

Gentner, D. and Bowdle, B. 2008. Metaphor as structure-mapping. In: Gibbs, R. W. (ed.) *The Cambridge Handbook of Metaphor and Thought.* Cambridge, England, U.K., Cambridge University Press.

Goatly, A. 1997. *The Language of Metaphor.* London, England, U.K., Routledge.

Haack, S. 1987–1988. Surprising noises: Rorty and Hesse on metaphor. *Proceedings of the Aristotelian Society*, 88, 293–301.

Hacking, I. 2002. *Historical Ontology.* Cambridge, Massachusetts, U.S.A., Harvard University Press.

Hajer, M. 2003. A frame in the fields: policymaking and the reinvention of politics. In: Hajer, M. and Wagenaar, H. (eds.) *Deliverative Policy Analysis: Understanding Governance in the Network Society.* Cambridge, England, U.K., Cambridge University Press.

Hansen, A. and Cox, R. 2015. *The Routledge Handbook of Environment and Communication.* Abingdon, England, U.K., Routledge.

Hausman, C. 2006. A revire of prominent theories of metaphor and metaphorical reference. *Semiotica*, 1, 213–230.

HC. 2010. *Proposals for Ireland's Landscapes.* Kilkenny, Co. Kilkenny, Ireland, The Heritage Council.

Hesse, M. 1980. *Revolutions and Reconstructions in the Philosophy of Science.* Brighton, England, U.K., The Harvester Press.

Knowles, M. and Moon, R. 2006. *Introducing Metaphor.* Oxford, England, U.K., Routledge.

Kohout, M. and Kopp, J. 2020. Green space ideas and practices in European cities. *Journal of Environmental Planning and Management*, 63, 2464–2483.

Kövecses, Z. 2002. *Metaphor: A Practical Introduction.* Oxford, England, U.K., Oxford University Press.

Lakoff, G. 1993. The contemporary theory of metaphor. In: Ortony, A. (ed.) *Metaphor and Thought.* Cambridge, England, U.K., Cambridge University Press.

Lakoff, G. and Johnson, M. 2003. *Metaphors We Live By.* London, England, U.K., University of Chicago Press.

Lemke, J. 1998. Multiplying meaning: visual and verbal semiotics in scientific text. In: Martin, J.R. and Veel, R. (eds) *Reading Science: Critical and Functional Perspectives on Discourses of Science.* New York City, New York, U.S.A., Routledge.

Lemos, N. 2007. *An Introduction to the Theory of Knowledge.* Cambridge, England, U.K., Cambridge University Press.

Lennon, M. 2015. Green infrastructure and planning policy: a critical assessment. *Local Environment,* 20, 957–980.

Lyon, G. 2000. Philosophical perspectives on metaphor. *Language Science,* 22, 137–153.

Matsler, A. M., Meerow, S., Mell, I. C. and Pavao-Zuckerman, M. A. 2021. A 'green' chameleon: exploring the many disciplinary definitions, goals, and forms of "green infrastructure". *Landscape and Urban Planning,* 214, 104145.

Maybin, J. 2001. Language, struggle and voice: the Bakhtin/Volosinov writings. In: Wetherell, M., Taylor, S. and Yates, S. J. (eds.) *Discourse Theory and Practice: A Reader.* London, England, U.K., SAGE Publications Ltd.

Medina, J. 2005. *Language: Key Concepts in Philosophy.* London, England, U.K., Continuum.

Medina, J. 2007. *Speaking from Elsewhere.* Albany, New York, U.S.A., State University of New York Press.

O'Neill, J. H.Holland, A. and Light, A. 2008. *Environmental Values.* Oxford, England, U.K., Routledge.

Rhodes, R. A. W. 2018. *Narrative Policy Analysis: Cases in Decentred Policy.* London, England, U.K., Palgrave Macmillan.

Richards, I. A. 1965[1936]. *The Philosophy of Rhetoric.* Oxford, England, U.K., Oxford University Press.

Ricoeur, P. 2002[1975]. *The Rule of Metaphor.* Oxford, England, U.K., Routledge.

Rydin, Y. 2003. *Conflict, Consensus, and Rationality in Environmental Planning: An Institutional Discourse Approach.* Oxford, England, U.K., Oxford University Press.

Schaffer, F. C. 2006. Ordinary language interviewing. In: Yanow, D. and Schartz-Shea, P. (eds.) *Interpretation and Method: Empirical Research Methods and the Interpretive Turn.* New York, New York, U.S.A.,M.E. Sharpe.

Schiappa, E. 2003. *Defining Reality.* Edwardsville, Illinois, U.S.A., Southern Illinois University Press.

Schön, D. A. 1993. Generative metaphor: a perspective on problem-setting in social policy. In: Ortony, A. (ed.) *Metaphor and Thought.* Cambridge, England, U.K., Cambridge University Press.

SDCC. 2010. *South Dublin County Development Plan 2010–2016.* Dublin, Ireland, South Dublin County Council.

Semino, E. 2008. *Metaphor in Discourse*, Cambridge, England, U.K., Cambridge University Press.

Semino, E. and Demjén, Z. 2016. *The Routledge Handbook of Metaphor and Language*. Abingdon, England, U.K., Taylor & Francis.

Shiff, R. 1978. Art and life: a metaphoric relationship. In: Sacks, S. (ed.) *On Metaphor*. Chicago, Illinois, U.S.A., University of Chicago Press.

Stern, J. 2008. Metaphor, semantics, and context. In: Gibbs, R. W. (ed.) *The Cambridge Handbook of Metaphor and Thought*. Cambridge, England, U.K., Cambridge University Press.

UF and IEEM2010. *Green Infrastructure: A Quality of Life Issue*. Dublin, Ireland, Urban Forum and the Institute of Ecology and Environmental Management.

Wagenaar, H. 2011. *Meaning in Action: Interpretation and Dialogue in Policy Analysis*, Armonk, New York, U.S.A., M.E. Sharpe.

Weaver, R. M. 1985. *The Ethics of Rhetoric*. Davis, California, U.S.A., Hermagoras Press.

Winch, P. 1958. *The Idea of a Social Science*. London, England, U.K., Routledge and Kegan Paul.

Wittgenstein, L. 1953. *Philosophical Investigations*. Oxford, England, U.K., Blackwell.

Yanow, D. 1996. *How Does a Policy Mean? Interpreting Policy and Organisational Actions*. Washington DC, U.S.A., Georgetown University Press.

Yanow, D. 2000. *Conducting Interpretive Policy Analysis*. London, England, U.K., Sage Publications.

Yee, A. 1996. The causal effects of ideas on policies. *International Organisation*, 50, 69–108.

5   Rationality Resonance

Practice Accord

The capacity of green infrastructure (GI) to resonate with the forms of rationality prevalent in planning practice appears rooted in its ability to discursively appeal to existing perspectives on the function of the planning profession and those allied to it. In addition, the expression's currency among such practitioners seemed to derive from its alignment with the epistemological assumptions manifested by common forms of disciplinary discussions, and the opinion that GI could be delivered via the conventional apparatus employed in policy formulation and direction. These practice-harmonising phenomena are respectively identified and discussed below as *good planning, language familiarity* and *existing planning vehicles.*

Good Planning

The connotative reasoning resultant from the interpretive requirements of GI's entitlement appears to have engendered the widespread opinion among those interviewed that 'the model for good green infrastructure is the model for good planning' (Interviewee A7). Indeed, many of those interviewed expressed the opinion that GI, as a form of infrastructure planning, should be integrated into planning practice. In this context, several interviewees insinuated that a GI-focused policy approach represented good planning practice. As commented by one local authority officer,

> ... is good planning not about, you know, figuring out what you have and documenting it; figuring out what you want and then figuring out a way to get from A to B, is that not good planning? Is that evidence-based planning; that's kind of what it's about, isn't

DOI: 10.4324/9781003413608-6

it? good planning has always been about that I think. So I don't really think green infrastructure is this new radical idea, I think it's just good planning in many ways, you know.

(Interviewee B20)

Here GI was equated with 'good planning' through an assumption that it constitutes 'evidenced-based planning' in presenting a useful means to 'figuring out' both what needs to be planned and the methods by which this can be achieved. Consequently, GI is not perceived as a 'new radical idea' but rather what planning practice 'has always been about'. Such assumptions imply a 'technical-rational model' (Owens et al., 2004, 1945) of planning practice wherein landuse governance decisions are made upon the impartial appraisal of 'scientific' (Interviewee A2) information rather than being influenced by non-quantifiable abstractions such as values or emotions. By virtue of connotatively-reasoned interpretations of GI as 'infrastructure', it was perceived to resonate with such a technical-rational model. Consequently, several other interviewees interpreted GI as harmonising with the objectives of planning practice. As noted by one senior planning official,

I think insofar as you know, amenities and the natural world, whatever it is, is a very important part of ... what we are, what places are, their character ... what they offer back to people, which is fundamentally what planning is about. I think Green Infrastructure ... does encompass all of those elements and is a very important part of what planning is about ...

(Interviewee C10)

This opinion that GI is 'fundamentally what planning is about' was reciprocally reinforced by, and gave discursive weight to, the aspects of professional discussions concerning GI. A discernible characteristic of this disciplinary dialect was its harmonisation with the established lexicon of professional practice; its 'language familiarity'.

Language Familiarity

As outlined in Chapter 2, the history in Ireland of equating networked approaches to planning for biodiversity with infrastructure, and GI in particular, extends back to the 2002 EPA study on ecological networks. Although following the production of this EPA study, the term GI did not re-emerge until 2008, the evolving language of green space planning during the interim period was increasingly characterised by the ever more frequent recurrence of vocabulary centred on the concept of 'networks'. An illustration of this trend was the 'green network' approach advocated in the Galway City Development Plan 2005–2011 that was adopted in January 2005. Building on a framework first presented in the previous Galway City Development Plan (1999–2005), this document proclaimed that such a 'green network' offers the means to combine and coordinate the protection of natural heritage areas and facilitate the provision of open space for recreational purposes. One of the primary methods promoted for the realisation of this network was the creation of greenways. This equation of the Council's 'green network' with a 'greenways approach' that 'form[s] connections between urban areas and the natural hinterland and link habitats' (GCC, 2005, Section 4.3), employed and merged terminology circulating in tangential discourses regarding the provision of transport facilities, recreational amenity and the conservation of biodiversity.

As with the case of Galway City Council, the Dublin City Council Development Plan 2005–2011 that was adopted in March 2005 similarly fused network-associated expressions and concepts from discourses on mobility and nature preservation when discussing the provision of open space. Here it was expected that such open space would,

> ... contribute to the development of *green chains* or *networks*, which allow for walking and cycling *and* facilitate biodiversity.
>
> (DCC, 2005, 84) [Emphasis added]

This networked-focused approach, and the language engendered by it, was echoed and extended in many subsequent documents issued by local, regional and national planning governance bodies and QUANGOs in the period prior to the emergence of a specific GI discourse in November 2008. The GI discourse that emerged in November 2008 during the Malahide Green Infrastructure Conference subsumed these networked focused discourses on biodiversity conservation, recreation provision and mobility planning, while concurrently equating green space with infrastructure in much the same manner as had been advocated six years previously by the EPA study on ecological networks (Tubridy and O Riain, 2002). In this sense, the GI discourse that materialised in 2008 combined the familiar language of networked green space planning with the linguistic and conceptual associations of traditionally-perceived infrastructure. From 2008, the emerging GI discourse increasingly subsumed and amalgamated an array of formerly separate narratives ranging from those centred on anthropocentric utility to those with an ecocentric focus on biodiversity conservation. This is illustrated by the position of one local authority planner, who in representing views frequently encountered during interviewing, stated,

> In my opinion it's [GI] the collective term that's used for *connecting* up areas of open space, maybe both informally and formally that are maybe home to various biodiversity forms and ecosystems but *connecting* up those spaces through maybe sustainable modes of transport like cycle *routes* and *walkways* so the green is probably more the areas of land, the woodlands, the biodiversity areas and then *the infrastructure is the connections* between those areas by say cycle *routes* and walking *routes* and preserving those ecosystems and joining them maybe together as well.
>
> (Interviewee B15) [Emphasis added]

Representative of perceptions common among interviewees, GI was here inferred to constitute the 'collective term' describing both open space (green) and the connections (infrastructure) between them. Additionally, the provision of 'cycle routes and walking routes' was seen as commensurate with 'preserving ... ecosystems and joining them together'. As such, the delivery of physical infrastructure for 'sustainable modes of transport' was conceived as a means to preserve ecosystems. Thus, the traditional partition of planning objectives directed at anthropocentric utility and ecocentric biodiversity conservation was removed. In its place was the fusing of these previously separated objectives so that the provision of facilitates to meet society's needs, such as those for 'sustainable modes of transport', was equated with 'preserving' ecosystems. The ease with which this process of reconceptualisation was advanced following the GI Conference of 2008 seems to have been facilitated by presumptions that 'infrastructure' inherently entails 'networks'. As noted by one planning authority interviewee,

> ... you've got the word 'infrastructure' in there and again it's a word that planners are familiar with, like 'road infrastructure', 'green infrastructure', a 'network', all the other types of 'infrastructure', so it's a word that they're familiar with.
>
> (Interviewee B3)

This *language familiarity* was remarked on by many of those questioned, with one local authority official commenting,

The thing that attracted me about it was that it made sense, it was a language that seemed to me to make sense to the likes of engineers and planners.

(Interviewee B20)

Therefore, the apparent simplicity in comprehending GI was enabled, assisted and manifested by 'language familiarity'.

Existing Planning Vehicles

Coupled with *connotative reasoning*, the phenomenon of *language familiarity* influenced the means proposed for the delivery of GI. Particularly, the linguistically-induced sense of acquaintance with GI's epistemology and solution propositions suggested to practitioners that the *existing planning vehicles* frequently deployed in formulating and presenting policy direction were the most appropriate for GI planning. In this context, it was asserted,

To be effective, green infrastructure thinking and strategies need to be integrated into local area plans, city and county development plans, and regional planning guidelines so that the full benefits of this approach can be realised.

(Clabby, 2009)

Consolidating this opinion was the connotatively-reasoned perception of GI planning as 'scientific in its nature' (Interviewee A2). Thus, GI planning was seen to entail a rational process centred on the use of quantitative survey and mapping methods in the collation and analysis of data. This was conveyed by one NGO planner when stating,

It's a matter of using what you need to do, matching it with what your evidence is on the ground and kind of developing your methodology business. It's like evidence-based planning. Then linking that through to your zoning of land, through your policies and your development plan, to actually develop a coherent strategy for it.

(Interviewee E4)

In this sense, GI was viewed as 'evidence-based planning' that links through development plan policies in a manner commensurate with an objective to develop a 'coherent strategy'. Consequently, it was believed that GI planning activities facilitated a transition from data through to policies and zoning designations in the same format as that conceived to operate within the 'technical-rational model' of landuse planning (Owens et al., 2004, 1945). This assumption that GI planning could and should be delivered via the existing vehicles of planning practice was conveyed by a local authority planner when addressing the issue of planning application assessment,

> ... development would have to adhere to the objectives of the plan where green infrastructure would filter right through the plan policies and objectives in terms of, you know, transport and natural heritage proposals and maps particularly, and the more information that's on a zoning map and the more layers that are visible and that you have to comply with, the easier it would be to enforce...
>
> (Interviewee B15)

This assumption that GI should be delivered through the existing vehicles of 'evidenced-based planning' in policy development and planning application assessment was prevalent in both interview and documentary data. Such existing planning vehicles, and the rationale on which they are based, reflects connotatively-reasoned assumptions on GI's scientific foundation manifested and reinforced by familiarity with the language of 'design', 'networks' and 'services' frequently deployed in GI discourses. Furthermore, such presumptions of objectively underpinned logic resonate with convictions of planning as an 'evidence-based' discipline grounded in the detached systematicity of expert practitioners. Thus, among planning and allied practitioners, the assumed scientific foundations of GI bequeathed to it a perceived knowledge legitimacy, and consequently 'enunciative advantage'.

Enunciative Advantage

Consequent on perceptions of GI as something 'scientific', and given convictions regarding planning as an 'evidence based' discipline, the ability to authoritatively pronounce on GI necessitated a capacity to discuss versions of the world perceived as objective, factual and impersonal. Therefore, the perceived veracity of GI knowledge claims required the effacing of apparent interest-motivation from the production and dissemination of information ascertained in analysing this independent reality. Such a concern surrounding the appearance of neutrality in the structuring and communication of knowledge claims has been termed 'stake inoculation' (Potter, 1996). The documentary and interview data suggests that cartography was central to this process. The data similarly indicates that those advocating GI endeavoured to bolster the legitimacy of their proclamations by comparison with what they identified as GI planning activities occurring in other countries. Also evident was the role played by quantification in facilitating the appearance of neutrality. The role of each of these processes in the meaning making of the GI policy approach are therefore examined below.

Inoculation by Mapping

As discussed above, the connotative reasoning inherent to GI's entitlement evoked a scientific semblance that prioritised the perceived rational planning processes associated with conventionally-conceived infrastructure. Resultant from such inferences was the pervasive assumption that a significant element of 'evidence-based' GI planning rested on conducting analyses and presenting conclusions through the medium of cartography. As noted by Comhar (the Irish Sustainable Development Council),

> The collection, mapping and analysis of data to arrive at a plan for development and management of natural areas, open space and related resources - is commonly recognised as the crux of Green Infrastructure planning.
>
> (Comhar, 2010e, 63)

This foregrounding of cartography in GI discourses may be traced to what MacEachren (1995) distinguishes as connotations of 'veracity' and 'integrity'. These are specified as the implications of temporal and attributive precision commonly associated with impressions of accuracy in mapping, and the presumption of impartiality in the activities of scientifically-schooled cartographers. As an activity intrinsically associated with planning's existing vehicles, both interview and documentary data indicate that it is such assumptions of cartographic fidelity with an objective reality that gave weight to mapping as the means to furnish the 'evidence base' in GI policy formulation. As asserted by the Irish Heritage Council,

> Green Infrastructure planning involves mapping existing Green Infrastructure resources, assessing future needs, and charting where improvements or enhancements can be made, and where new Green Infrastructure can be provided in the future. Strategies

are evidence-based and generally use Geographical Information Systems (GIS) to collate, map and analyse information.

(HC, 2010, 24)

As such, cartography was viewed as a means to 'control' (Pickles, 2004) the organisation and provision of GI, and thereby the configuration of future spaces (Harley, 2001; MacEachren, 1995; Monmonier, 1991). Thus, rather than a neutral communicator of information, mapping GI can be interpreted as a process that 'selectively brings into being a world that is socially constructed' (Wood, 1992, 20) with its own affordances and constraints.

The perceived scientific legitimacy embodied in mapping was alluded to in several interviews when referenced as the primary mechanism to accurately analyse quantitative data and present it in a means conducive to facilitating a rational process of policy formulation. As noted by one consultant planner involved in the production of GI documentation,

> Well evidence in this case is obviously proper mapping, proper survey, proper mapping of the various elements which go into, into the resource, which is as we say, the natural biodiversity, the amenity, the cultural aspects, all of those things, that's very important as the evidence base: surveying it, mapping it and capturing it and then on that basis, then you proceed forward and make decisions on that. So it shouldn't be basically policy or ideas that come basically 'shooting from the hip'; it needs to be chased back into proper planning process.
>
> (Interviewee A10)

In its presentation as the scientific 'evidence base' for legitimate planning activities, the medium of communication was thereby intimately connected with the message it communicated (Yanow, 2000). Here the grounding of GI planning in cartography had 'a dimension of symbolic realism' (Harley, 1992, 241) in which the perceived impartiality of scientific assessment was implied. As such, cartography enabled map authors to legitimately proclaim the 'facts' of a situation from an advantageous enunciative position via appeal to the seeming objectivity engendered by stake inoculation in a 'will to truth' (Foucault, 1976, 55) of a desired reality. Put simply, maps legitimated that which was enunciated.

Legitimate Enunciation via Cartographic Presentation

As discussed above, the phenomenon of language familiarity brought into play by GI's *entitlement* stimulated connotatively reasoned assumptions of GI planning as the mapping and provision of green spaces to facilitate the maintenance of infrastructure servicing the development requirements of society while simultaneously assisting the conservation of biodiversity. In this sense, it was asserted that,

> Green infrastructure provides a wide range of invaluable ecosystem services and human quality of life benefits including:
> - biodiversity management and enhancement
> - water management including drainage and flood attenuation, filtration and pollution control
> - recreation and tourism
> - visual amenity and sense of place
> - sustainable mobility
> - food, timber and other primary production
> - regulation of micro-climates (green lung) and, potentially, climate change adaptation
>
> (UF and IEEM, 2010, 2)

By mapping areas to facilitate the planning and provision of this array of perceived GI functions, the scientific legitimacy afforded to cartographic activities engendered the apparent rational interpretation of anthropocentrically-orientated green space development as concurrently facilitating environmental conservation. Maps were thus employed as powerful tools in the generation of desired landuse functions wherein they effected actualisation of the objective facts constituting legitimate spatial realities (Corner, 1999, 225). Consequent on this wide range of functions attributed to GI, those who advocated it as a planning approach frequently employed cartography as a tool to construct a reality of functional coexistence within spaces by encompassing multiple landuses beneath the rubric of 'GI'. This was reflected in the work of the Comhar GI Consultant Team in formulating the Comhar document titled, 'Creating Green Infrastructure for Ireland: enhancing natural capital for human well being', published in August 2010.

As part of the document production process, Comhar and its GI Consultant Team organised a GI workshop, which was attended by an invited selection of identified stakeholders drawn from central state institutions, local and regional planning authorities, QUANGOs and NGOs (Comhar, 2010e). At this event a number of GI maps were presented by the consultants to the invited multi-disciplinary audience and feedback was requested. The function of these maps was to demonstrate the workings and benefits of a potential methodology for the collation of data, its cartographic expression, analysis and use for GI planning. Whereas the 'rational methodology' was favourably received by the audience, the content of the maps were questioned by a number of ecologists working for Dublin City Council. This was due to the identification on the GI maps as 'recreational & quality of life', lands[1] popularly used for recreational purposes but designated under European Union law for nature conservation as both a Special Protection Area and a Special Area of Conservation. Whereas 'recreational & quality of life' appeared an appropriate categorisation for the consultants (Interviewee A4), it was feared by the council's ecologists that categorising these lands as 'recreational & quality of life' on these GI maps would sanction intensification of their use for recreation and thereby threaten their ecological integrity (Interviewee B5). This instance of rupture in the conceptual fixing of landuses by way of cartographic labels indicates the perceived power of maps, and map categories in particular, in constructing the meanings that embody the authority to shape reality. Resolving this issue in a manner that maintained the perceived integrity of the GI concept entailed a phenomenon universally characteristic to Irish GI discourses, namely the dissolution of unifunctional landuse categories. Here the specification of landuse categories for single landuse purposes was revised to facilitate multiple landuses on the same site. This was effected through the inclusion of landuses within multiple landuse categories.

Thus, following this workshop, the maps produced and tabled for comment by the consultants were reviewed, and where deemed necessary, they were updated. It was agreed that rather than presenting one map indicating the area as 'recreational & quality of life', numerous 'layers' graphically portraying the variety of perceived GI purposes would have to be provided as separate maps for the same area to avoid the potential misinterpretation of GI as either unifunctional or of prioritising any one landuse above others. These maps were then assembled into a final multifunctional GI map representing the many uses of GI deemed commensurate for the area. In this way, the perceived priority given to 'recreational & quality of life' landuses was reduced, yet such landuses were not removed. Thus, both 'recreational & quality of life' and nature conservation landuses were accommodated on the site, without resolving concerns regarding the incompatibility of biodiversity conservation with the intensified use of the lands for recreational purposes. Indeed, following these revisions, the previous disagreement regarding spatial functions did not materialise in Comhar's work.

As such, in responding to contentions provoked by the perceived authority of map categories, new spatial typologies were engendered.

Within these new typologies, multiple landuses previously deemed incompatible were reconstituted as concordant via the presentational techniques and perceived scientific legitimacy of modern cartography. In this way, mapping methods effected the actualisation of new spatial realities to facilitate consensus and dispel potential disagreement surrounding GI's flexible signification and consequent latitude for application. Thus, cartography served as the means to legitimately enunciate on, and thereby constitute, the apparent objective reality of spaces.

Inoculation by Comparison

Another prominent stake inoculating mechanism employed by GI advocates was comparison. Central to this was managing the relationship between the identity of those referencing a knowledge claim, those identified as producing such a claim, and that upon which the claim was made. The stake inoculating potentials and properties of such relationships were explored by Erving Goffman and elucidated in his theory of *footing* (1981), wherein a threefold typology of reference is theorised. These are namely the *principal*, whose position the piece of speech is supposed to represent; the *author*, who does the scripting; and the *animator*, who says the words. These distinctions may be employed to exert influence on the appearance of neutrality as they can position the animator as 'just passing something on' (Potter, 1996, 143) – in this case, that which the author has produced regarding the principal. Indeed, 'it is through the paraphernalia of footing that speakers managed their personal or institutional accountability' (Potter, 1996, 122).

The role played by footing can be observed in the prevalence of comparison in discourses on GI planning in Ireland. Here a salient feature of such discourses was evaluation of the perceived condition of Irish GI planning relative to that of other jurisdictions. Such comparisons were employed to provoke action on the *principal* of innovative green space planning by the seemingly objective identification of progressive planning practices identified as widespread in other jurisdictions yet still absent in Ireland. These practices were subsequently referenced as models for how GI planning should be conducted in Ireland. In this sense, comparison was utilised to facilitate stake inoculation via footing in articulations by GI advocates (*animators*) who referenced external cases deemed non-partisan to Irish planning debates. Thus, comparison was exercised as a way of generating an apparent distance between the potentially conceived partisan agendas of GI advocates and the 'facts' of a situation as stated by unprejudiced independent *authors*.

Legitimate Enunciation via Comparison

The importance of footing in the stake inoculation that facilitated the apparent impartiality required for *enunciation advantage*, was postulated by several interviewees and expressed by one local authority officer who concluded,

> One advantage I found in trying to do something new or different is if you can show that another country has done it and what they've used the information for, then it can be very valuable.
>
> (Interviewee B3)

Thus, many advocates of GI (*animators*) stressed the long history and widespread adoption of approaches to green space planning (*principal*) in countries thought to possess advanced landuse planning systems (*authors*). As publicly proclaimed by one advocate,

> Since the 1990s, green infrastructure approaches to planning and managing green space have been developing in the USA and, more recently, in the UK where Natural England – the Government's advisor on the natural environment – has been promoting the concept. In continental Europe, 'green structure' planning has long been a feature of city planning, for example in Copenhagen, and – in recent decades – ecological networks have been planned and developed in several countries.
>
> (Clabby, 2009)

Hence, there was an implication that Irish practitioners may consult the efforts of foreign planning practice in devising indigenous green infrastructure planning approaches. Additionally, listing the progress made by other planning systems with regards to GI planning implied that Irish planning practice was falling behind that of other progressive systems. Furthermore, it is noted that included beneath the rubric of GI in this statement are 'approaches to planning and managing green space' in the U.S.A. and the U.K., 'green structure' planning in continental European countries, and 'ecological networks' in several unspecified nations. Indeed, resultant from the connotative reasoning and the flexible signification inherent to the interpretation of GI, a recurring feature of Irish planning policy discourses was that they referenced a variety of readings as to both what landuse functions GI refers to, and the spatial applicability of the approach. This polysemy was consequently reflected in the diversity of identified and referenced

GI activities promoted as offering models for green space planning (*principal*) in Ireland. In seeking the *enunciative advantage* bequeathed by perceptions of objectivity, those advocating (*animators*) the application of such exemplars portrayed the assumed necessity of *stake inoculation* by furnishing the citation of particular examples (*authors*) detailing where such planning approaches have been applied. As stressed by one QUANGO ecologist,

> ... if you have to justify different measures you're taking, then you can say well, you know this is in line with the green infrastructure developments as reflected in Holland, wherever the hell it is, the States, and you know, people go oh that's interesting.
>
> (Interviewee C7)

Accordingly, both interview and documentary data indicate that those advocating (*animators*) different interpretations of green space planning (*principal*), referenced different examples (*authors*) of GI activities dependant on the specific comprehension of GI that they were forwarding. A feature of Irish GI advocacy was therefore the use of footing to achieve stake inoculation in the promotion of specific perspectives on green space planning by bestowing on such perspectives the legitimacy of apparent impartiality demanded by practitioner self-assessment of planning as an 'evidence-based' discipline. This phenomenon enabled the presumed legitimacy and simultaneous advocacy by multiple parties of different understandings of what GI entails. Such assorted interpretations facilitated, and were facilitated by, reference to a variety of diverse examples of activities seen to constitute progressive GI practices. In referencing these identifiable cases (*authors*), the promoters (*animators*) of GI offered an interpretation of what they deemed to be its relevance for green space planning (*principal*). These approaches generally resonated with their personal and/or professional biases, be that for health, transport, conservation or a range of other possible functions.

Inoculation by Quantification

Although less prevalent than the prominent roles occupied by mapping and comparison as means to analyse, represent and advocate GI, also evident in many policy documents and interviews were references to numerical data and the processes of quantification. Underpinning such references was the connotatively-reasoned comparability of GI with conventionally-conceived infrastructure wherein quantitative methodologies were thought inherent to its delivery. As noted by one planner involved in the production of GI documentation,

> [i]t's [GI] looking at open space resources as we would grey infra-structure. We have a piece of land, a resource, what do we want it to do. How much of that do we want it to do. So you plan and design for that and then you can measure its performance.
>
> (Interviewee A2)

Indeed, Aronowitz (1988) outlines how the authority endowed by the scientific semblance of such quantification is predicated on the con-flation of 'knowledge' with 'truth'. This influence on the production of 'truth effects' (Foucault, 1972[1969]) is characterised by deference to the assumed integrity of quantification as a means by which to accu-rately represent reality. For as noted by Kingdon (1984, 98), quantified information 'acquires a power of its own that is unmatched by issues that are less countable'. Thus, statistics may be employed as a way to legitimise knowledge claims that convey meaning seemingly indepen-dent of those who employ them, and thereby facilitate enunciative advantage.

Legitimate Enunciation via Quantification

The enunciative advantage endowed by stake inoculation via quanti-fication may help conceal the normative impetus of counting activ-ities by force of appeal to the perceived objective methodologies of scientific measurement. Hence, the deed of measuring may imply 'a need for action, because we do not measure things except when we want to change our behaviour in response to them' (Stone, 1997, 167). The process of quantification itself may thus serve as a tacit message signifying that something is of a sufficient magnitude to warrant numerical investigation, and therefore should be taken ser-iously. It is in this sense that some of those interviewed suggested that a cost-benefit analysis of GI's merits may carry greater weight than reference to normative arguments. As noted by one local authority official,

> I would like to see the debate started on the basis of cost-benefit as opposed to on the basis of some sort of feel good kind of approach. I think it would be good to see a fairly rigorous approach adapted in terms of cost-benefit.
>
> (Interviewee B21)

Accordingly, as discussed above with reference to cartography, acts of quantification can assume a metaphorical character that support both the perceived importance of something and the objectivity of its assessment (Throgmorton, 1993). In this context, and with reference to GI, one interviewee stressed that,

> [u]ntil you can come up with a method of actually quantifying it, and mapping and quantifying it and making it real, then they're just concepts, you know, they're not that meaningful for people.
>
> (Interviewee C8)

The legitimating and issue-highlighting functions of counting were ardently forwarded by certain parties to the GI advocacy discourse and can be observed in the endeavours of Comhar to present GI as an objectively assessed economic benefit. Playing a central role in the advocacy of a GI planning approach in Ireland, arguments for GI advanced by Comhar were closely aligned with a discourse focused on the 'monetarisation' of 'natural assets'. This was illustrated by the director of Comhar in his presentation of an economics centred argument for the introduction of multifunctional GI planning at the Irish Planning Institute's Annual Conference in April 2010 (Comhar, 2010c). Such an endorsement of a cost-benefit argument for the adoption of GI planning was sustained by Comhar in its presentation at the Parks Professional Network Seminar Day in June 2010 (Comhar, 2010d), when it was announced that the estimated worth to Ireland of the ecosystems services delivered by GI was €2.6 billion. In the same month, Comhar hosted a workshop on The Economics of Ecosystems and Biodiversity (Comhar, 2010a). This workshop involved a plenary session wherein a series of presentations were provided outlining the economic worth of biodiversity and the methodologies that can be employed in its valuation. With a focus on an economic assessment of GI's value, the published report recommended as a priority the,

> [i]dentification, quantitatively and qualitatively of the economic and social benefits of ecosystem services delivered by Green Infrastructure in monetary terms and also the social gains to health and quality of life.
>
> (Comhar, 2010b, 23)

In such instances, counting the value of GI may be seen as a means to remove it from possible associations with ex-ante value rationalities (Kornov and Thissen, 2000; Owens et al., 2004) and foreground a mathematically-determined instrumental rationality for its introduction. Here, a positivist repertoire grounded in numeracy was employed to present arguments as founded on externalised facts by 'divesting agency from fact constructors and investing it in facts' (Potter, 1996, 158). In doing so, a stake inoculation of those 'facts' was achieved simultaneous to conveying the important story about which 'the facts speak for themselves'. The particular 'facts' of the GI approach advanced by those who advocated its adoption, was that GI planning policy is a scientifically-identified cost-effective means to solve a multitude of problematic issues and deliver numerous benefits to society. It was in such circumstances that normatively founded proclamations on what was believed to be requisite action obtained the *enunciative advantage* of scientific legitimacy by the seemingly objective 'evidence base' upon which planning was viewed to operate. As outlined in Chapter 3, such legitimately-enunciated normative imperatives centred on the perceived need to give greater weight to green space issues in planning policy formulation.

Functional Advantage

As discussed in the previous chapter, resultant from the connotative reasoning of GI as analogous to conventionally-conceived infrastructure, those advocating this approach both assumed and asserted its servicing functions. However, the latitude for interpretation bestowed by the term's flexible signification elicited numerous possibilities for the expression's application. Thus, rather than representing a clearly defined and unifunctional application, the GI approach was seen as validly and concurrently pertaining to a broad assortment of planning issues. In this way, most of its advocates stressed multifunctionality as a key advantage of the approach. Promoters of GI often foregrounded this inferred benefit in literature seeking to advance the approach's practical merits. This is illustrated by one such document when declaring,

> GI is multifunctional at every scale, for example in considering water basin management, the opportunity for habitat creation and enhancement should also be exploited. Green solutions to hard issues such as flooding, coastal erosion and carbon sequestration should be considered first as an alternative to expensive grey infrastructure. All environments have potential to restore biodiversity and this can be enhanced with GI planning. GI projects generate tourism and employment dividends by improving access

to existing natural assets and opening up new recreational and leisure opportunities.

(UF and IEEM, 2010, 4)

Exhibited in this document and prevalent throughout interview and documentary data, was that through presumptions rooted in ontologies derived via connotative reasoning, GI was viewed by most of its advocates as an inherently 'networked' planning approach that may be 'planned', 'designed', 'delivered' and 'managed' (KCC, 2011, Chp. 14, 19) 'at every scale' (UF and IEEM, 2010, 4). This perception was reinforced by extrapolations induced by language familiarity and the extension of antecedent tangential discourses. Consequently, those promoting GI married this conceptual assumption with the perceived advantage of multifunctional potential in pronouncing the approach's capacity for the effective spatial integration of geographically isolated and functionally disparate areas. This popular opinion was advanced by one QUANGO official when claiming,

> ...it [GI] has a number of functions: it can function as a sort of recreation, sort of transport link, it can function as [sic] biodiversity network allowing species and things, plants and animals, species and things to move, including ourselves actually.
>
> (Interviewee C3)

Several of those interviewed postulated that such a networked approach to the provision of various services necessitated amendments to policy guidance hierarchies in catering for the multifunctional potential promoted by GI. In this, many conjectured that the functional advantage of the GI approach stimulates innovative landuse planning protocols that require an ability to straddle the traditionally discrete administration of services provision. As proposed by one consultant planner involved in the production of GI documentation,

> [i]t's [GI] multifaceted ... it's seeking to group a number of objectives under the one title, that's probably the key element of it. I don't see anything very new about any of the aspects of it ... they're all addressed in more detail in their various subsets; SUDS, water management, recreation, landscape, they're all very well addressed within their own disciplines ... the innovation if you like or the uniqueness of it may be that it's being grouped, you know, as a number of objectives within an overall strategy or something like that.
>
> (Interviewee A10)

Emanating from this assessment was the conclusion that in compelling innovative and 'evidence-based' policy approaches to cater for the multifunctional potential of green spaces, GI planning represents an aggregation of normally disparately-planned issues in a fashion that renders their respective merits easier to convey. Those advocating this approach subsequently argued that the matters encompassed by GI thereby enjoyed greater appeal in planning policy formulation. In this context, one local authority officer noted,

> [t]hat to me is a good thing about the green infrastructure thing, that you're not talking to people about ten agendas, you're talking to them about one, even though it might encompass six things underneath it, but at least it's one thing. So you're not asking them to do biodiversity, archaeology, architecture, landscape, water; you have it kind of packaged and its maybe easier then for people to kind of get a grip on that in their thinking.
>
> (Interviewee B20)

Indeed, several of those promoting the GI approach emphasised the role it plays in facilitating greater weight of consideration to numerous issues commonly perceived as neglected in policy formulation. This they claimed is achieved by both assembling such issues for presentation in an easy-to-understand format and bestowing on them a sense of import often lacking in their assessment. GI may thereby be conceived as a strategy to place various policy issues on the policy decision agenda so that they would receive 'more of a hearing' (Interviewee A10). This was achieved by providing clarity regarding the problematic ambiguity surrounding numerous policy issues, such as how to ensure effective flood management, biodiversity protection, landscape conservation, as well as sustainable transport and recreational amenity provision. This capacity 'to give a simple message' (Interviewee B16) was enabled by the connotative reasoning of GI as a form of infrastructure that can be planned, designed, delivered and managed using the same methods as conventionally-conceived 'grey' infrastructure. This reasoning prompted assumptions of apparent simplicity regarding GI's comprehension, which consequently induced perceptions of clarity on issues of ambiguity. Nevertheless, the degree of flexible signification facilitated latitude for the interpretation and subsequent application of GI for a variety of purposes and to an array of spatial typologies. This permitted those advocating a GI approach to advance the concept as one which facilitates multifunctionality. In doing so, GI presented 'a badge to join up a whole range of ideas' (Interviewee B16) in which a variety of varying policy issues were 'kind of packaged' (Interviewee B20) 'as a way of selling [the] concept' (Interviewee B16). In this way, the 'strength in numbers' (Interviewee A10) presented by the construal of GI as a planning approach performed policy work by: (a) communicating the importance of certain issues; (b) outlining the benefits of their consideration; (c) providing direction for landuse governance; and (d) placing normally neglected issues on the decision agenda.

Summary and Conclusions

The discussion provided in this chapter extends that offered in Chapter 4 by outlining how the meanings induced by the interpretive features of *connotative reasoning, apparent simplicity* and *flexible signification* prompted comprehensions of GI that resonated with the prevailing rationality of Irish planning practice. Specifically, perceptions of GI as analogous to conventionally-conceived infrastructure prompted assumptions of GI planning as congruent with *good planning* in the 'planning, management and engineering of green spaces and ecosystems in order to provide specific benefits to society' (UF and IEEM, 2010, 2). This view was fortified by the evolving discourse's *language familiarity* and its consequent subsuming of antecedent narratives concerning planning, engineering and conservation. Buttressing this were connotatively-reasoned assumptions that the *existing planning vehicles* employed in normal landuse policy formulation were appropriate to the constitution and implementation of GI guidance.

A disciplinary self-assessment of planning as an 'evidence-based' activity provided the backdrop profiling interpretations of 'infrastructure' as that which is designed, delivered and managed via scientific methods. This enabled those who advocated green 'infrastructure' to stress its legitimacy as an objective and systematic approach to green space planning. Consequently, GI was seen to offer *enunciative advantage* through impartial assessment and conclusion specification permitted by *cartographical presentation* and *quantification*. Such enunciative advantage was braced by the *comparison* of Irish endeavours with the application of GI in other jurisdictions by parties unconnected to planning debates in Ireland. Hence, GI resonated with practitioner presuppositions regarding the objectives and technical-rational methods of planning practice by virtue of its interpretation through the prism of the prevailing rationality operative within the arena of landuse governance.

Resultant from the *apparent simplicity* and *flexible signification* engendered by the conceptual constitution of GI via *connotative reasoning,* those who advocated this planning approach stressed the benefits it presented by way of its *multifunctional potential*. Here promoters of GI emphasised the role it could play in facilitating the integration of areas commonly lamented as functionally divergent or geographically isolated. The approach's supporters espoused its capacity to advance multiple issues, including those heretofore largely disregarded. In so doing, GI was seen to endow issues perceived as normally neglected with greater weight of consideration in policy formulation through associating them with issues enjoying greater attention. Thus, it was through perceptions of *practice accord, enunciative advantage* and *functional advantage* that GI acquired *rationality resonance* among planning practitioners and allied professionals.

Illustrated in Figure 5.1 is the reciprocal relationship between the 'naming effects' of GI's entitlement (see Chapter 4) and the elements of *rationality resonance* discussed throughout this chapter. Graphically portrayed is how the relationship between the particular characteristics

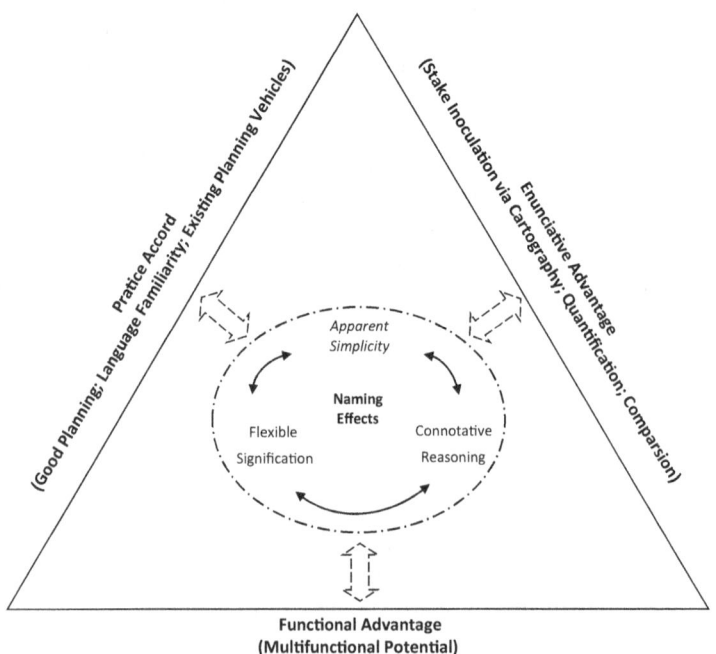

Figure 5.1 Diagrammatic Representation of the Relationships between the *Naming Effects* of GI's *Entitlement* and *Rationality Resonance*
Source: author

associated with the interpretation of GI's meaning facilitated its resonance with the prevailing rationality of Irish planning practice. This in turn influenced how the GI concept was interpreted and represented by those advocating its use in planning policy formulation. By examining the perceived resonance of GI with the prevailing rationality of Irish planning practice, this chapter furnishes a base upon which to investigate how different meanings were framed and advanced by different parties seeking to promote a GI planning approach. Thus, it is to this issue that we turn to in the next chapter.

Note

1   Bull Island, Dublin City.

References

Aronowitz, S. 1988. *Science as Power: Discourse and Ideology in Modern Society*. Minneapolis, Minnesota, U.S.A., University of Minnesota Press.

Clabby, G. 2009. *Green Infrastructure: Critical Infrastructure for a Smart Economy* [Online]. Dublin, Ireland, Comhar SDC. [Accessed 12.11.10].

Comhar. 2010a. *Workshop on The Economics of Ecosystems and Biodiversity.* Dublin, Ireland.

Comhar. 2010b. *Creating Green Infrastructure for Ireland: Enhancing Natural Capital for Human Well Being*. Dublin, Ireland, Comhar SDC.

Comhar. 2010c. Green New Deal & Green Infrastructure. *Irish Planning Institute Annual Conference.* Tullamore, Offaly, Ireland.

Comhar. 2010d. PPN Presentation: Green Infrastructure Enhancing Natural Capital for Human Wellbeing. *Parks Professional Network Seminar Day.* Dublin, Ireland.

Comhar. 2010e. *Workshop on a Green Infrastructure Strategy for Ireland*. 8 February 2010. Dublin, Ireland.

DCC. 2005. *Dublin City Development Plan 2005–2011*. Dublin, Ireland, Dublin City Council.

Foucault, M. 1976. *The Will to Knowledge: The History of Sexuality: Volume 1*. London, England, U.K., Penguin Books Ltd.

Foucault, M. 1972[1969]. *The Archæology of Knowledge*. London, U.K., Routledge.

GCC. 2005. *Galway City Development Plan 2005–2011*. Galway, Ireland, Galway City Council.

Goffman, E. 1981. *Forms of Talk*. Oxford, England, U.K., Basil Blackwell.

HC. 2010. *Proposals for Ireland's Landscapes*. Kilkenny, Co. Kilkenny, Ireland, The Heritage Council.

Harley, J. B. 1992. Deconstructing the map. In: Barnes, T. J. and Duncan, J. S. (eds.) *Writing Worlds: Discourse, Text and Metaphor in the Representation of Landscape*. London, England, U.K.: Routledge.

Harley, J. B. 2001. *The New Nature of Maps.* Baltimore, Maryland, U.S.A., John Hopkins University Press.

Kingdon, J. W. 1984. *Agendas, Alternatives and Public Politics.* New York City, New York, U.S.A., Harper Collins.

Kornov, L. and Thissen, W. A. H. 2000. Rationality in decision- and policy-making: implications for strategic environmental assessment. *Impact Assessment and Project Appraisal*, 18, 191–200.

MacEachren, A. M. 1995. *How Maps Work: Representations, Visulaizations and Design.* New York, U.S.A., Guildford Press.

Monmonier, M. 1991. *How To Lie With Maps.* Chicago, U.S.A., University of Chicago Press.

Owens, S., Rayner, T. and Bina, O. 2004. New agendas for appraisal: reflections on theory, practice, and research. *Environment and Planning A*, 36, 1943–1959.

Pickles, J. 2004. *A History of Spaces: Cartogaphic Reason, Mapping and the Geo-Coded World.* Oxford, England, U.K., Routledge.

Potter, J. 1996. *Representing Reality.* London, England, U.K., SAGE Publications Ltd.

Throgmorton, J. A. 1993. Survey research as rhetorical trope: electric power planning arguments in Chicago. In: Fischer, F. and Forester, J. (eds.) *The Argumentative Turn in Policy Analysis and Planning.* London, England, U. K., Duke University Press.

Tubridy, M. and O Riain, G. 2002. *Preliminary Study of the Needs Associated with a National Ecological Network.* Wexford, Ireland, Environmental Protection Agency.

UF and IEEM. 2010. *Green Infrastructure: A Quality of Life Issue.* Dublin, Ireland, Urban Forum and the Institute of Ecology and Environmental Management.

Wood, D. 1992. *The Power of Maps.* New York, U.S.A., The Guildford Press.

6 Narrative Modality

The GI Discourse Coalition

Discursive Affinities and Contamination

Academics who advance a discourse analysis approach to the study of planning and environmental issues suggest that what gives traction to specific ideas in the policy process is ascription to a particular series of narratives that clarify meaning in situations of policy ambiguity (Epstein, 2008; Roe, 1994; Stone, 2002; Griggs and Howarth, 2019; Valdez et al., 2018). The discourse coalitions that emerge from the endorsement by various agents of these narratives facilitate the perception of shared allegiance to a specific policy solution while concurrently enabling a multitude of interpretations of the meaning of that policy solution. As noted by Hajer,

> [w]hat unifies these coalitions and what gives them their political power is the fact that its actors group around specific story-lines that they employ whilst engaging in environmental politics. It can be shown that although these actors might share a specific set of story-lines, they might nevertheless interpret the meaning of these story-lines rather differently and might each have their own parti-cular interests.
>
> (Hajer, 1995, 13)

Nevertheless, unrestricted individual license of interpretation is impli-citly constrained by the discursive format and content of the narratives. Consequently, narratives cluster possible interpretations of meaning and position the actors who ascribe to them into coalitions of broadly similar, albeit not necessarily identical, interpretations. As discussed in Chapter 3, the broad problem narrative giving rise to green infrastructure (GI)

DOI: 10.4324/9781003413608-7

advocacy was the perception that the multitude of issues related to green spaces were often assigned low priority in the planning system. This broad problem narrative begot perceptions of a shared predicament among a wide spectrum of dissatisfied parties whose concerns were perceived to be fundamentally related to green space planning. As previously discussed in Chapter 4 regarding *flexible signification* and the *functional expectations* of green spaces, and in Chapter 5 with respect to *language familiarity* and *multifunctional potential*, such interests in green space planning may be diverse. In this sense, the power of GI to assist the emergence of a discourse coalition may be understood as derived from its ability to facilitate 'discourse affinities' (Hajer, 1995, 66) among the varying issue-specific narratives of those parties advocating the importance of green space planning. Hajer proposes that such affinities may not refer to actors and their intentions but rather 'operationalizes the influence of discursive formats on the construction of problems' (Hajer, 1995, 67). Such problem construction was discussed in Chapter 3 when examining the promotion of GI as a solution narrative to address the perceived low profile allocated to green space planning by endowing such areas, and the issues seen as associated with them, greater weight of consideration in the policy formulation process. Consequently, in perceiving GI as offering a means to raise the profile of green space issues in planning activities, those with varying motives for promoting green space consideration in policy development formed a discourse coalition centred on a 'narrative of necessity' in the advocacy of GI (see Chapter 3). Hajer (1995) theorises that in the case of a particularly strong affinity, discursive elements not only resemble one another, but an exchange of terms or concepts may exist. He terms such an occurrence 'discursive contamination' (Hajer, 1995, 67). Indeed, discourses on GI were replete with instances of discourse contamination during the period of the concept's emergence in Ireland. This is illustrated in the conclusion of one local authority officer when postulating the benefit of the term GI,

> I suppose green in people's minds is now synonymous with ecological or like nature and then infrastructure, I suppose ... if you're dealing with engineers, they very much think of the roads and the rail and that kind of infrastructure, so I suppose if you're presenting nature in that context then maybe it helps that understanding so yeah, I think it's a good description.
>
> (Interviewee B19)

As discussed in Chapters 3 and 4, the discursive weight alluded to here was prompted by connotations of 'necessity' with the word 'infrastructure'.

This facilitated the perceived compatibility of GI with numerous discourses centred on the provision of services to society. Additionally, Chapter 4 discussed how associations stimulated by the word 'infrastructure' simultaneously suggested that the provision of GI was a rational planning activity that could be undertaken by employing the familiar policy formulation and implementation tools conventionally deployed in planning practice. This interviewee also suggested that the word 'green' was perceived to relate to an environmentally-sensitive approach to human activities. Furthermore, as discussed in Chapters 4 and 5, the word 'green' was concurrently seen to relate to a wide range of both 'formal and informal spaces' (Interviewee B15). The conjunction of the words 'green' and 'infrastructure' in the expression 'green infrastructure' thereby stimulated connotations of GI planning as a rationally-conceived, necessary and environmentally-sensitive approach to green space planning that could be delivered through the existing scientifically grounded policy vehicles commonly employed in planning practice, such as cartography and quantitative assessment (see Chapter 5). Consequently, perceptions on the utility and inclusivity of the GI narrative fostered the formation of a broadly encompassing discourse coalition wherein the manifold and potentially incompatible interests of various parties could co-exist by virtue of their 'discursive affinity' with the perceived importance of green space planning. This phenomenon was recognised by one QUANGO interviewee when reflecting,

> [s]o depending on where your interest originally starts, you know people will take a primary interest in one aspect but appreciate and almost latch on to the other aspects as a way of selling the idea. So in that way it's [GI] a sort of useful term … it allows a lot of people who have overlapping interests to come together and sort of share the space.
>
> (Interviewee C3)

This supposed capacity to suspend potential differences in forwarding a narrative from which all parties to the discourse coalition were perceived to benefit accords with the theoretical application of 'myth' by Yanow (1996) in the context of policy and organisational analysis.

Suspension

In her interpretive analysis of the evolution of Israeli community centres, Yanow employs the term 'myth' to indicate a 'narrative created and believed by a group of people which diverts attention from a puzzling part of their reality' (Yanow, 1996, 191). Drawing from anthropological studies and echoing the work of Roland Barthes (2009), the concept of myth advanced here is not conceived as an assessment of a narrative's veracity as myths are neither true nor false in the empiricist sense. Rather, perception of their truthfulness is dependent on ascribing to them. As such, myth in the context of policy analysis refers to a particular narrative format that facilitates ascription by a broad spectrum of issue-specific interests through providing apparent commensurability in situations where plausible discrepancies may coincide. Myths achieve this through suspending conflict by 'masking the tensions between or among incommensurable values' (Yanow, 2000, 80) and deflecting attention away from potential logical inconsistencies or possible incompatibilities in that which is enunciated. In this way, myths facilitate *narrative modality* by enabling multiple parties with various interests to espouse a particular narrative consequent on its perceived benefit for the specific concerns they seek to advance. As noted by one local authority officer, 'I suppose there's scope for us all, we can all have a chunk of it [GI] and there's a benefit to us all' (Interviewee B23). This mythic quality is particularly germane to the GI narrative in light of how the term's latitude for interpretation resulted in GI's application to a variety of issues normally considered discrete (see Chapter 4). Despite this wide array of issues, those who advocated a GI planning approach demonstrated the narrative's mythic property through the suspension of potential conflict in presuming general consensus regarding the term's meaning. Thus, as surmised by one QUANGO interviewee who advocated a GI approach to planning,

> I think at the moment it's probably generally a simple enough concept...I'd imagine there would be a certain amount of consensus on what it's about.
>
> (Interviewee C5)

Such assumed general consensus facilitated the suspension of potential conflict through the supposed co-existence of multiple interests within the GI discourse coalition. This was achieved via common ascription to the concept of landuse multifunctionality which was deemed a central advantage of the approach (see Chapter 5). Consequently, most of those interviewed considered that a GI approach to planning enabled the commensurable and simultaneous utility of lands for a variety of purposes. Such conjecture was not solely confined to inferred compatibilities in recreational space provision and biodiversity protection. Rather, these suppositions extended to a broad array of issues perceived as encompassed by the expression 'GI'. Indeed, the endorsement of GI as simultaneously providing numerous and contiguous functions was a pervasive view promulgated by those ascribing to the precepts of the approach. Thus, the attested multifunctional potential of landuses avowed by this planning approach signified to those who propounded it that,

> ... [GI] has a number of functions; it can function as a sort of recreation sort of transport link, it can function as biodiversity network allowing species and things, plants and animals, species and things to move, including ourselves actually. And also that from a heritage point, a cultural heritage point of view, it's also a way you can look after perhaps heritage infrastructure such as disused railway lines or even things like stone walls or old roads.
>
> (Interviewee C3)

Possible tensions in the GI discourse coalition consequent on varying potential landuse incompatibilities were held in suspension by the proposition of GI's capacity to effect functional reciprocity. This phenomenon was illustrated by one local authority planner in outlining the perceived advantages of a multifunctional-focused GI approach to urban drainage,

> ... one of the things we're exploring, because it's quite an obvious one, is if you can get something like the SUDS[1], a large attenuation pond area into open space, just it's a good example of the benefits that can be achieved throughout green infrastructure in terms of its open space. It helps the open space, it helps SUDS, so it helps water and then you can get wildlife within it. So it helps the whole wildlife, so there's three or four different areas which has a positive impact. You know so that's a good example where green infrastructure can be a success ...
>
> (Interviewee B24)

In addition to promoting GI planning for the concurrent realisation of several benefits to society and nature, the view asserted here also presents the GI approach as furnishing the physical conditions in which the provision of these gains is mutually reinforcing. Such reasoning alleviated prospective discord within the GI narrative by suspending potential views on landuse incompatibilities and thereby assisting consolidation of the GI discourse coalition. The force of this logic led many of those interviewed to conclude that there are no clearly identifiable disadvantages to the GI planning approach. This opinion was expressed from many quarters, with one NGO planner stating,

[d]isadvantages to the actual approach. Let me think. It's hard to see, it's hard to see any specific disadvantage of it.

(Interviewee E4)

Thus, the GI narrative neutralised possible differences of opinion through the suspension of potential logical inconsistencies and landuse incompatibilities via appeals to the shared advocacy benefits of a multifunctional and synergistic approach to green space planning. Consequently, negative evaluation of the approach's possible disadvantages were deflated and the consolidation of the GI discourse coalition effected. However, not all feasible disagreement was suspended by reference solely to synergistic multifunctionality and the advantages of shared advocacy. Where anticipated dissonance remained, a process of 'deflection' manifested.

Deflection

Deflection differs from *suspension* in that it is specifically directed at averting identified potential criticisms of a narrative's logical consistency. Rather than avoiding discussion of possible incompatibilities, as is the case with suspension, the process of *deflection* engages discernible prospective discord. While comparatively less evident than suspension in the context of GI advocacy in Ireland, instances of deflection may be observed in attempts to pre-empt concerns regarding the compatibility of recreational provision and biodiversity protection. Here an appeal to 'balance' in issue assessment and planning was employed to deflect potential criticisms. As surmised by one local authority planner, and echoed in the assertions of many other interviewees,

> [w]e need to find a fine *balance* between development on the one hand and preservation of amenities and heritage assets on the other, so it's a *balance* between the two.
>
> (Interviewee B15) [Emphasis added]

Supporting this requirement for balance, some GI advocates who addressed the issue conjectured that the GI approach may be the best means to facilitate evaluation, representation and arbitration of potentially competing interests in green space planning activities. This view was conveyed by a consultant involved in the production of GI advocacy documentation when surmising,

Bull Island[2] is a good example of an area where, there's maybe *competing interests*, and some people are going to say, well listen, maybe I might shut the whole place off, put up the fences and keep it for biodiversity only … But at the same time, without something like green infrastructure, well then the sort of biodiversity elements of open spaces and parks, some of the institutional lands and so forth, I mean that just doesn't get a look in. So on *balance* I think it's very much to the benefit of biodiversity.

(Interviewee A4) [Emphasis added]

Here potential concern regarding landuse compatibility was deflected. This was achieved while concurrently advocating the deployment of GI as a means to facilitate landuse multifunctionality in the conservation of biodiversity. In this sense, the appeal to 'balance' forwarded by some of those seeking to deflect criticism of GI planning entailed the repositioning of biodiversity protection from ecocentric approaches focused on the exclusion of human activities to perceptions that the human use of habitats can be an important means for their conservation. As discussed in Chapter 3, this representation of such areas in terms of anthropocentric utility was perceived as a requirement in fomenting the 'narrative of necessity' thought important in addressing the opined low profile of green spaces in planning policy formulation.

As such, the processes of *discursive affinities* and *contamination*, in conjunction with *suspension* and *deflection*, may be identified as enabling the cohesion of a GI discourse coalition. However, given the polysemy of the term 'GI', and the varying interests which it was seen to address, appreciation of how GI sustained such a broad constellation of concerns necessitates an exploration of how the GI narrative offered a unifying solution to the policy predicaments of multiple parties. An identification and examination of such mechanisms reveals a subtle process wherein the naming effects prompted by the term GI facilitated 'narrative modality'. Here GI emerged as an inclusive solution narrative that assimilated the problem narratives of numerous other planning discourses. Thus, building upon the discussion presented above, the analysis now turns to how such *narrative modality* operated.

Narrative Modality

In discussing the significant influence metaphor may play on problem setting in social policy, Schön (1993) outlines how the connotative reasoning[3] engendered by metaphor may help induce perceived solutions to problems which otherwise lack resolution. Schön also suggests that a metaphor may furnish language for the communication of a problem-solution narrative where lucidity of articulation was seen as previously absent. With regard to GI, both the presentation of a solution and the 'rebranding' (Interviewee B3) of existing problem–solution narratives was given latitude of application resultant from the 'discursive affinities' and 'contamination' (Hajer, 1993, 1995, 2005) prompted by the term's flexible signification. As noted by Hajer,

> ... metaphors provide a common ground between various dis-courses. Actors are thus given the opportunity to create their own understanding of the problem, re-interpreting various elements of knowledge outside their specific realm of competence ...
>
> (Hajer, 1995, 62)

In this sense, as a metaphor, GI served as a vehicle,

> for the discursive reduction of complexity, allowing people to communicate over complex policy issues.
>
> (Hajer, 2003, 105)

Consequently, those seeking solutions to unsolved problems felt com-fortable adopting the GI narrative as a resolving discourse to their particular issue(s) of problematic policy ambiguity. In addition, those desiring to successfully communicate the merits of their planning activities felt comfortable applying the GI discourse to their existing problem–solution narratives as a means to achieve greater weight of consideration for their specific narratives in policy formulation. In such contexts, a process of *narrative volunteering* could be said to have occurred.

Subsequent to the process of *narrative volunteering*, those employing GI for their own problem resolution purposes felt comfortable con-solidating the potency of the GI narrative by exploiting the term's flexible signification and apparent simplicity in assigning it as a solu-tion to other issues beyond the concerns of their original problematic issue or problem–solution narrative. Capacitated by discursive affinities and contamination, in conjunction with the processes of suspension and deflection, those seeking to advance the influence of the GI nar-rative, and consequently their own issue specific interests, advocated the attachment of the GI narrative to other problematic issues which they perceived as lacking a coherent solution narrative. Additionally, such advocates sought to reframe as GI existing problem–solution narratives circulating in other policy arenas. Resultant from assumptions of GI's scientific grounding, together with the apparent simplicity and flexible signification prompted by the term's interpretation (see Chapter 4), those engaged in such advocacy behaviour did so with regard to policy issues normally considered beyond their professional competences. In such contexts, a process of *narrative application* could be said to have occurred.

By means of such *narrative volunteering* and *narrative application* activities, those advocating the GI narrative conceived it as justifiably incorporating, representing, and where necessary resolving, the problems and problem–solution narratives circulating in a broad array of policy arenas. Consequently, the scope of issues addressed by the GI narrative was perceived to expand with a corresponding increase in the size of the GI discourse coalition. In this sense, GI achieved *narrative modality*. A tracing of this process is presented below.

Narrative Volunteering

Solution Adoption

Several of those interviewed suggested that the GI narrative furnished a solution for the problematic issues they were endeavouring to address. In attaching the GI solution narrative to their problem, they conveyed its benefit in generating a coherent problem–solution narrative previously seen as absent in attempts to resolve a particular issue. While the content of these issues varied, in such cases of *solution adoption*, the problematic issues addressed by GI were usually directly associated with the area of expertise or role of the professional[4] seeking such resolution. This phenomenon was observed and expressed by one consultant involved in the production of GI documentation when concluding on perceptions of GI's merits relative to the concern of landscape professionals and ecologists,

> [p]eople who are involved in landscape certainly, people who are involved in ecology certainly would see, okay, here is a concept that provides the opportunity to enshrine our particular area into the planning system where it hasn't been previously. I know ecologists would certainly make the comment that ecological planning and biodiversity planning is very piecemeal and reactive. In the past, here was a site, protect it, at all costs, draw a circle around it. Landscape the same, you know. Plant a tree to make something look prettier if you can, or whatever … green infrastructure is

basically offering a potential ideal solution to those problems ... and that's why it's appealing to people.

(Interviewee A2)

This perceived capacity of GI to serve as a solution to problems associated with green space planning was not confined to the activities of landscape professionals and ecologists, but was also conceived as applicable to planning problems centred on the management of built environments. As noted by one local authority planner,

[i]t [GI] would attempt to address failures in the past where we've planned for residential areas in towns and cities, that lacked open space and that lacked recreational areas ... it's about linkages between areas because good urban design says that areas should be connected and we should be able to walk ... and that you can provide green infrastructure to link areas and they can serve other purposes such as, you know, sustainable transport routes.

(Interviewee B17)

In this sense, GI was perceived to simultaneously address three interconnected problems experienced in the development management[5] of new residential areas. Specifically, the interviewee suggested that GI not only addresses problematic issues regarding recreational open space provision but in doing so could concurrently facilitate pedestrian permeability in residential areas and consequently furnish opportunities for non-motorised transport. This opinion was reflected in the opinions of several town planners interviewed, with most advancing the view that GI offered the potential for greater attention to green space planning issues in development proposals submitted for consideration to local planning authorities. As commented by one local authority planner when appraising the perceived problem of inadequate open space provision and its poor configuration in development proposals,

... generally stuff tends to come into authorities pre-determined almost. You know that way like, that it would be planned on the basis of where the road's coming in and then they won't deal with all the other issues joined up. They'll just deal with them and then just kind of shoehorn them in around everything else. So it's [GI] to try and say no, this [GI] is actually a centre stage piece ... you have to think about this before you design what you're doing.

(Interviewee B16)

Thus, through the process of *solution adoption*, particular issues previously deemed unresolved by current planning procedures were considered remedied via a GI approach. This was achieved by the attachment of GI as a solution narrative to a specific problem narrative where an identifiable means of problem remedy was formerly seen as absent. In this way, GI facilitated *narrative modality* among a cohort of professionals seeking to address specific problems encountered in their practice activities. Augmenting this process was a parallel exercise of 'elective rebranding' wherein those seeking to profitably communicate the merits of their planning activities elected to 'rebrand' their existing problem–solution narratives as GI in an effort to achieve greater weight of consideration for these narratives in policy formulation.

Elective Rebranding

Coupled with the apparent discursive affinities of multiple issue-specific narratives, many interviewees alluded to a desire to 'rebrand' their existing problem–solution narratives as GI. This they believed would lend them greater weight in the policy development process. As noted by one local authority official involved in GI advocacy,

> I think it's [GI] kind of a broad idea ... so to me, my bottom line is any gain from the ideas that I'm interested in is a gain, and if it's delivered through green infrastructure, great!
>
> (Interviewee B20)

Such reasoning suggests a perceived unproblematic advantage in the *elective rebranding* of planning activities as GI in endeavouring to bequeath such activities greater weight of consideration in policy deliberations. This opinion was conveyed by one QUANGO planner when explaining the decision to employ the term GI in green space planning guidance directed at local authority planners,

> I suppose we used the term because it's probably, you know, to a certain extent it has a cache at the moment ... I suppose the document is aimed a lot at planners and local authorities and we know the term would resonate with them. They would know what we are talking about when we spoke about green infrastructure so we used it almost as a code word.
>
> (Interviewee C5)

Here the elective rebranding of advocated planning activities as GI was opined to endow such pursuits with greater significance in green space planning policy formulation. Underpinning this elective rebranding was the perceived resonance of GI with traditional modes of planning practice. As noted by one planner engaged in GI advocacy,

> I think it [GI] is an appropriate term ... on a practical level you're never going to get it through, you're never going to get this accepted by your engineers, by your county managers ... unless you start to think about it in the way that other issues are considered ... your roads programmes or your rail programmes or your procurement programmes, all this type of stuff. Unless you start thinking about it in the systemic way that other things are thought about, you're never going to get it up there at [sic] the agenda. Actually think of it as this is infrastructure; this is important stuff, this is capital ... I think that for purely for those reasons, I think it's an appropriate term.
>
> (Interviewee E4)

The advocacy and perceived advantage of such elective rebranding was enabled by GI's flexible signification and the discursive affinities. These attributes allowed it to share a conceptual space with other narratives seeking to address green space-related issues. Consequently, this led many of those interviewed to conclude that GI was simply a contemporary and potentially profitable rebranding of activities already extant in planning practice. This was related by one local authority interviewee when reflecting,

> ... the council had been developing these walkways within the city so then we came along after that and kind of said, well actually lads, do you know what, they're actually 'green infrastructure'. So it kind of was the reverse way around than maybe the model would suggest.
>
> (Interviewee B7)

As discussed in Chapters 4 and 5, the naming effects engendered by the interpretation of what GI meant fomented perceptions of GI's resonance with the prevailing rationality of planning practice. These naming effects subsequently prompted perceptions of relevance and potential benefit via the discursive affinities of numerous issue-specific green space narratives. This consequently assisted GI's *narrative modality* by stimulating shared opinions regarding the profile-raising prospect offered by GI's communicative potential. Accordingly, advocates of particular approaches to the management of specific green space issues engaged in the *elective rebranding* of their existing problem–solution narratives to benefit from the discerned traction of GI in deliberations on planning policy formulation. As a result, the GI discourse coalition expanded both in member composition and the content of issues addressed. The processes of *elective rebranding* and *solution adoption* are illustrated in Figure 6.1.

Narrative Application

As a counterpart to *narrative volunteering*, a parallel process of *narrative application* occurred. This facilitated increased frequency of use of the term GI by a multitude of different agents with respect to an array of issues. In essence, *narrative volunteering* encompasses problem solving while *narrative application* concerns solution advocacy. Specifically, *narrative application* involves the appropriation of unresolved problem narratives circulating in non-heretofore related policy discourses and attaching an idea or concept as a solution to the issues referenced therein. It also entails the imposed rebranding as that idea or concept of existing problem–solution narratives and activities not termed that idea or concept by those advancing them. A detailed explanation of these processes is provided below.

Problem Appropriation

In the case of Ireland's GI story, *problem appropriation* involved the requisition by GI advocates of a problem narrative which they were

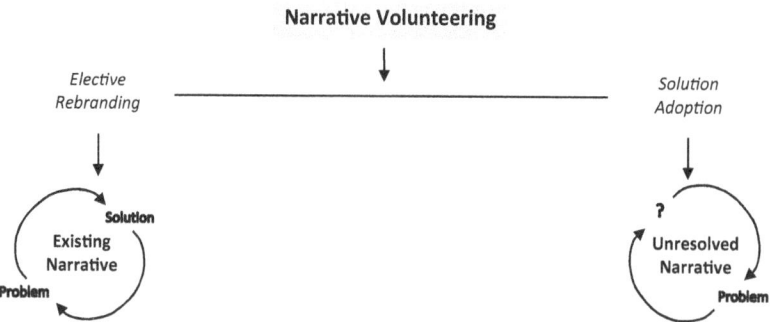

Figure 6.1 Elective Rebranding & Solution Adoption in 'Narrative Volunteering'
Source: author

not previously considered to be involved in resolving. Such advocates subsequently specified GI as a resolution to the problematic issue referenced. This was facilitated by GI's flexible signification and its consequent latitude for application, wherein those advocating GI as a planning approach advance it as proffering a resolution to an array of potential problems. Such a process was alluded to by several interviewees and exemplified in many documents. For example, this phenomenon is illustrated by a 'Comhar Commentary' when it was asserting that,

> [a] growing body of evidence underlines the many health benefits of green infrastructure. Well-designed, attractive and safe green spaces are important as places to exercise. Green spaces provide play areas for children and have positive benefits for community mental health … Green infrastructure also provides many economic benefits … High-quality green infrastructure translates into higher property values and rents. It helps to attract and to hold on to the high-value industries, entrepreneurs and workers needed to underpin the knowledge economy …
>
> (Clabby, 2009)

Here GI is forwarded as addressing a broad spectrum of issues ranging from physical and mental health, through to economic development and property values. Furthermore, this article continues by advancing GI's capacity to tackle flood management and pollution control, facilitate mitigation of the urban heat island effect, as well as enabling climate change adaptation while concurrently assisting in meeting the requirements of environmental legislation. Such problem appropriation and the ensuing advocacy of GI as a remedy to the perceived concerns of several problem narratives was often articulated in ambiguous terms, with the lack of specificity seemingly counterbalanced by the legitimacy endowed by GI's perceived resonance with the prevailing rationality of planning practice (see Chapter 5). Thus, GI was presented as a means to address a multitude of issues. Resultant from its *rationality resonance*, it was perceived that applying a GI approach to remedying such issues 'makes sense'. Consequently, those advocates of GI profited from GI's perceived capacity to endow issues with greater weight of consideration in planning policy formulation. This process thereby facilitated expansion of the GI *discourse coalition* and added momentum to its *narrative modality* by legitimating the confident and flexible application of GI as a solution narrative via *problem appropriation*.

Imposed Rebranding

Paralleling the process of *problem appropriation* was the rebranding of existing problem–solution narratives as GI by advocates not normally party to the resolution of the issues referenced. Therefore, this process differs from the process of *'elective' rebranding* in that it was conducted by those not normally party to discussions on the issue in question. As such, it is termed *'imposed' rebranding*. Instances of this process were relatively common in discourses concerning GI, with several of those interviewed rebranding the activities of others as GI. These rebranded activities often varied widely. For example, one consultant planner who promoted a GI approach in urban design cited campaigners for urban gardening as advocating GI when asserting,

> ... they're campaigners, they're doing what they do ... they're like ... 'get out of my way I want this to happen', but it is Green Infrastructure ... this is what it's about, guerrilla gardening.
>
> (Interviewee F1)

Another consultant involved in the production of GI strategies referenced as GI the more conformist activity of designing integrated constructed wetlands (ICW),

> ... the ICW is maybe the, it is the flagship green infrastructure project really, isn't it, because it links wastewater treatment to biodiversity and visual amenity, possibly even, you know, is compatible with recreational green space.
>
> (Interviewee A7)

Such imposed rebranding was also evident in the context of guidance endorsement by GI advocates and may be illustrated by reference to the Comhar document entitled 'Green Infrastructure for Ireland' when it states,

> The European Council of Spatial Planners, in a document titled 'Try it This Way: Checklist for Sustainable Development at the Local Level', reiterates the importance of Green Infrastructure planning in urban areas (although without naming it as such), suggesting possible components of the urban Green Infrastructure network and stressing the importance of its connection to the urban hinterland.
>
> (Comhar, 2010, 16) [Parenthesis in original]

Although less prevalent in application, of greater effect appeared to be the imposed rebranding of existing statutory plans as exhibiting a GI approach. This is illustrated by the relatively frequent reference made to the Loughmacask Local Area Plan (LAP) as a model of GI planning. Exemplifying this process of *imposed rebranding* was the declaration of the GI advocacy document entitled 'Green Infrastructure: a quality of life issue',

> [t]he LAP [Loughmacask Local Area Plan] demonstrates a clear understanding of context that informed the design and layout of the plan. It is evident that this Plan embraces the concept of considering Green Infrastructure from first principles in the preparation of an LAP and that Green Infrastructure sits comfortably within the plan making process. The Green Infrastructure of this scheme is multifunctional and is not a burden on the public purse but rather a common sense approach to providing the environmental services for a new urban community.
>
> (UF and IEEM, 2010, 12)

However, the Loughmacask Local Area Plan (KKCC, 2008) did not actually mention GI and was produced prior to the re-emergence of the GI discourse in Ireland in November 2008. This rebranding of the Loughmacask Local Area Plan as an exemplar of GI planning was a recurrent feature of interviews and was illustrated by one local authority planner when seeking to reference the emergence of GI within their area of jurisdiction. In doing so, this planner noted citation of the same local area plan as GI at the annual Irish Planning Institute conference of 2011,

> I'm not quite sure when green infrastructure started coming into play here … I'd say the Loughmacask plan. It's like I said, it was used in the planning conference by the guy who was doing the presentation as an example of green infrastructure. Green infrastructure was never actually mentioned once in the plan, but the policies are written to favour that sort of set up.
>
> (Interviewee B13)

Thus, although acknowledging that GI was not referenced in the Loughmacask Local Area Plan, those advocating a GI approach to planning cited it as an example of GI via the process of *imposed rebranding*. In this manner, GI's flexible signification facilitated the imposed rebranding of other discourses perceived as having discursive affinities with the green space planning concerns of GI advocates. Consequently, the composition of the GI discourse coalition was viewed as expanding in parallel with the increasing range of issues embraced by the GI narrative. As a result, GI's *narrative modality* was further enhanced as it was perceived to legitimately provide a solution narrative to a growing number of problematic green space planning issues. The processes of *problem appropriation* and *imposed rebranding* are illustrated on Figure 6.2.

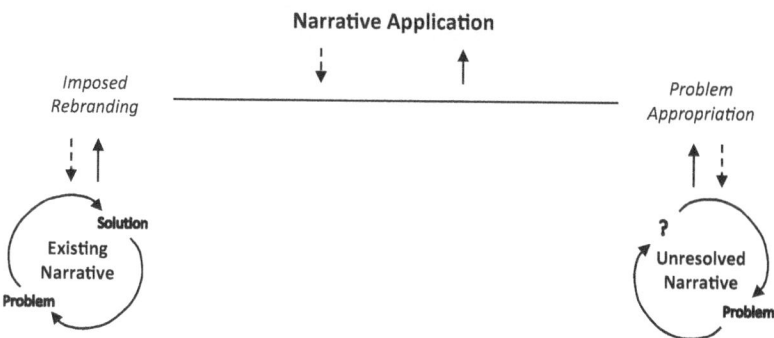

Figure 6.2 Imposed Rebranding and *Problem Appropriation* in 'Narrative Application'

Source: author

Summary and Conclusions

This chapter outlined how GI's flexible signification permitted attachment of the GI narrative to a multiplicity of problematic issues. This was undertaken by introducing and explaining how the 'mythic' qualities of *suspension* and *deflection* operated in facilitating the emergence and evolution of a discourse coalition converging on perceived discursive affinities concerning green space planning policy formulation. Subsequently discussed was how these discerned commonalities were buttressed by 'discursive contamination' resultant from the interpretation of GI's meaning(s) relative to conventionally-conceived 'infrastructure'. Succeeding this, it was shown that perceived resonance with the prevailing rationality of planning practice fortified the GI discourse coalition by affording the GI narrative apparent legitimacy. Demonstrated was how in combination, these processes prompted and enabled GI advocacy. This chapter identified, described and explained a process whereby once agents adopted GI as an issue-specific solution or electively rebranded their narrative as GI (problem solving), they subsequently sought to apply the

GI narrative to other discourses (solution advocacy). This process is graphically summarised in Figure 6.3. It is hypothesised that through the processes of *narrative volunteering* and *narrative application*, both the membership of the GI discourse coalition and the content of the issues encompassed by the GI narrative were augmented. In this way, the *narrative modality* of GI was facilitated.

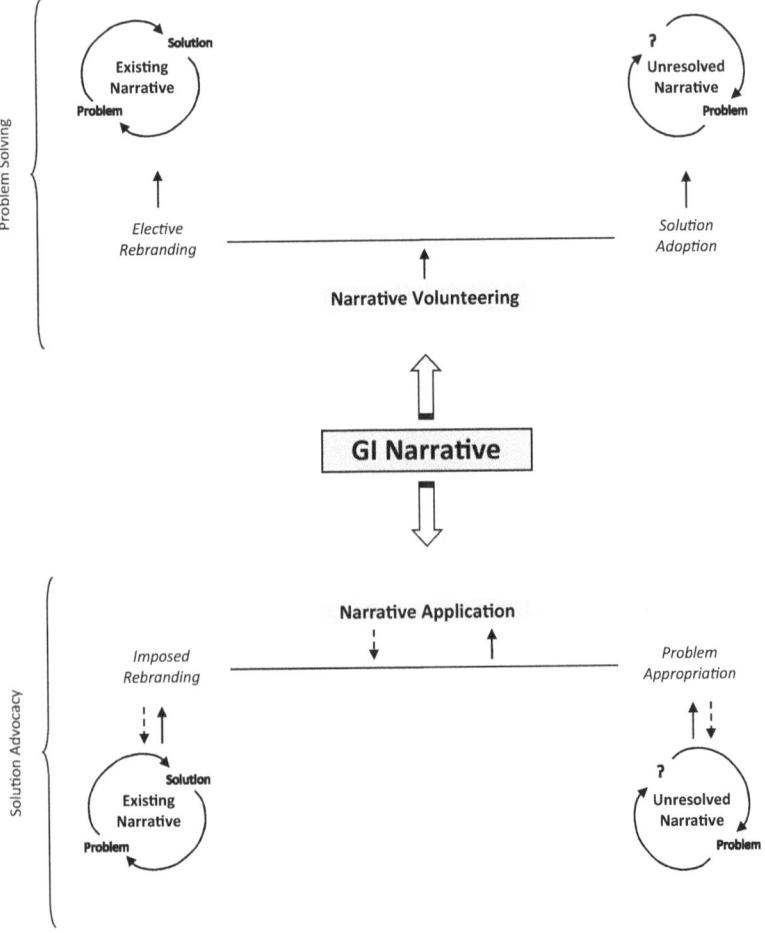

Figure 6.3 Narrative Volunteering and Narrative Application
Source: author

Notes

1 SUDS is the acronym used for Sustainable Urban Drainage Systems. This is the context in which it is used by this interviewee. However, it can be written as SuDS where it is intended to refer to the less 'urban' focused engineering concept of Sustainable Drainage Systems.

2 Located in Dublin City and previously discussed in Chapter 5 with respect to the role of cartography in producing new spatial realities.

3 Schön (1993) refers to the idea-constituting properties of 'generative metaphors'.

4 All of those interviewed occupied what would normally be considered 'professional' positions, with their organisational functions requiring the possession of, at minimum, a primary degree. Most of those interviewed possessed postgraduate degrees, with several of those interviewed having doctorates.

5 'Development management' in Irish planning practice refers to the activity of evaluating planning applications against adopted policy to arrive at a decision to grant permission, refuse permission or grant permission subject to certain conditions.

References

Barthes, R. 2009. *Mythologies*. London, England, U.K., Vintage.

Clabby, G. 2009. *Green Infrastructure: Critical Infrastructure for a Smart Economy* [Online]. Dublin, Ireland, Comhar SDC. [Accessed 12.11.10].

Comhar. 2010. *Creating Green Infrastructure for Ireland: Enhancing Natural Capital for Human Well Being*. Dublin, Ireland, Comhar SDC.

Epstein, C. 2008. *The Power of Words in International Relations: Birth of an Anti-Whaling Discourse*. London, England, U.K., The MIT Press.

Griggs, S. and Howarth, D. 2019. Discourse, policy and the environment: hegemony, statements and the analysis of UK airport expansion. *Journal of Environmental Policy & Planning*, 21, 464–478.

Hajer, M. 1993. Discourse coalitions and institutionalization of practice: the case of acid rain in Britain. In: Fischer, F. and Forrest, J. (eds.) *The Argumentative Turn in Policy Analysis and Planning*. London, England, U.K., Duke University Press.

Hajer, M. 1995. *The Politics of Environmental Discourse: Ecological Modernisation and the Policy Process*. Oxford, England, U.K., Oxford University Press.

Hajer, M. 2003. A frame in the fields: policymaking and the reinvention of politics. *In:*Hajer, M. and Wagenaar, H. (eds.) *Deliberative Policy Analysis: Understanding Governance in the Network Society*. Cambridge, England, U.K., Cambridge University Press.

Hajer, M. 2005. Coalitions, practices, and meaning in environmental politics: from acid rain to BSE. In: Howarth, D. and Torfing, J. (eds.) *Discourse Theory in European Politics: Identity, Policy and Governance*. Basingstoke, England, U.K., Palgrave MacMillan.

KKCC. 2008. *Loughmacask Local Area Plan*. Kilkenny, Co. Kilkenny, Ireland, Kilkenny County Council.

Roe, E. 1994. *Narrative Policy Analysis: Theory and Practice*. London, England, U.K., Duke University Press.

Schön, D. A. 1993. Generative metaphor: a perspective on problem-setting in social policy. In: Ortony, A. (ed.) *Metaphor and Thought*. Cambridge, England, U.K., Cambridge University Press.

Stone, D. 2002. *Policy Paradox: The Art of Political Decision Making*. New York City, New York, U.S.A., W.W. Norton.

UF and IEEM. 2010. *Green Infrastructure: A Quality of Life Issue*. Dublin, Ireland, Urban Forum and the Institute of Ecology and Environmental Management.

Valdez, A.-M., Cook, M. and Potter, S. 2018. Roadmaps to utopia: tales of the smart city. *Urban Studies*, 55, 3385–3403.

Yanow, D. 1996. *How Does a Policy Mean? Interpreting Policy and Organisational Actions*. Washington DC, U.S.A., Georgetown University Press.

7 Policy Entitlement

Theorising the Policy Process

As discussed in this book's Introduction, policy formulation is conventionally conceived as a form of applied problem-solving wherein a 'technically orientated rational model of policy making' (Fischer and Gottweis, 2012, 2) is perceived to operate. In this view, where difficulties arise in formulating solutions, these are seen to be rectified by more information about the problem at hand (Fischer, 2003; Hill and Varone, 2021). However, such a linear comprehension of the policy process fails to account for how agents resolve issues of problematic policy ambiguity where there exists a 'state of having many ways of thinking about the same circumstances or phenomena' (Feldman, 1989, 5). In such situations a conventional understanding of policy making fails as problem identification is rendered inconclusive and solution formulation is left indecisive. Consequently, in such circumstances, studying the policy process requires attention to the persuasive power of representation in providing clarity and direction on problem recognition and solution specification (Goodin et al., 2006; Stone, 2012; Rhodes, 2018). Accordingly, as argued in Chapter 1, enhancing knowledge of the policy process requires consideration of how reality is represented in policy debates through interpretations of signification, significance and applicability. Thus, it necessitates concentration on how 'meaning making' functions both in and through the policy process.

This attention to meaning making involves an acknowledgement that the reality of a policy entails an 'perceptual interpretive element' (Kingdon, 1984, 115) wherein meaning is produced and 'situated in a particular context' (Yanow, 2006, 228). Emerging during the early 1990s, literature centred on this 'interpretive turn' (Yanow, 2007, 405) to policy analysis argues that it is through such meaning making processes that representations of reality are constructed, and the persuasive work of policy gets

DOI: 10.4324/9781003413608-8

done. However, as discussed in Chapter 1, most labour within the field of interpretive policy analysis assumes the policy process to be a site of struggle for dominance and control over issues of contested meaning. Even where such adversarialism is not foregrounded, its existence is pre-supposed by advancing post-positivist methods of conflict prevention, most commonly in the form of normatively-orientated 'collaborative planning' (Healey, 2005, 2012) or 'deliberative practice' (Buchstein and Jorke, 2012; Dryzek, 2012).

Myerson and Rydin (1996) adopt a different approach by con-centrating on how rhetoric functions to realise the purposes for which it is deployed rather than focusing on how deliberation should be con-ducted to facilitate better policy. However, their study centres on the contours of an argument's structure wherein the rhetorical composition is explicit. Consequently, the presuppositions of interlocutors that may give currency to certain forms of argumentation are not investigated in detail. This leaves aside 'how' the interpretation of rhetorical devices assists or inhibits traction in policy debates. Additionally, their study does not address 'how' the new understandings engendered by rheto-rical devices may be disseminated. Furthermore, they do not address 'whose' interests are served by such rhetorical devices.

Thus, while the work of Myerson and Rydin is laudable in its focus and approach, there remains a dearth of interpretive policy analysis dedicated to investigating how meaning making operates through pro-cesses of policy persuasion in the absence of disagreement or explicit attempts to avoid conflict. Yanow (1996) goes some way to addressing this deficit with regard to her employment of a theory of 'myth' in explaining how the use of metaphor may suspend and deflect logical inconsistencies and potential criticism in policy debates. While she places meaning making centre stage in policy formulation, the illumi-nating possibilities of this work are restricted by not detailing how the resolution of problematic policy ambiguity may result from a counter-intuitive process whereby solutions are specified in advance of problem identification and subsequently applied to a range of problematic policy issues. Therefore, although she furnishes another commendable example of interpretive policy analysis, gaps in our knowledge remain regarding the interpretation, traction and dissemination of new policy concepts. Specifically, the following key gaps are evident in theoretical explanations of the policy process:

a   Policy process theory fails to adequately account for how agents resolve issues of problematic policy ambiguity through widespread voluntary and unanimous support across a range of organisation

types and disciplines for a policy proposal where there exists significant potential for dispute.

b   Theorists have largely failed to identify and comprehensively explain
the interpretive mechanisms that engender new policy concepts.

c   Theories of the policy process have neglected to describe and
explain how meaning making in and through the policy process may
prompt new perceptions of reality that render problematic policy
ambiguity amenable to resolution by existing policy formulation
practices.

It is contended that these gaps in our knowledge can be addressed by
innovatively harnessing a discourse analysis approach to the interpretation of policy process dynamics. Such a theoretical approach enables the
formulation of an explanation of how a new policy approach may
emerge, evolve and be adopted by successfully balancing the requirement for theoretical abstraction with sensitivity to the context in which
policy activity is situated. By deduction from the study of green infrastructure's (GI) emergence and evolution in Ireland between November
2008 and November 2011, the understanding of the policy process
dynamics supplied below stresses the importance of particular forms of
representation to the ascension of a new policy approach. Focused on
the ontological, epistemological and coalition-stimulating consequences
of the strategic use of naming, this hypothesis has been termed a 'policy
entitlement'.

Policy Entitlement

The term 'entitlement' is borrowed from the work of Burke (1966) who, in
stressing the selective and abstractive function of naming, draws attention
to how this procedure may concurrently abbreviate the complex while
specifying the ontological status of something. A rhetorical consequence
of such 'entitlement' is that it prompts the impression that what is entitled
has always existed autonomous of its entitlement and was waiting to be
discovered as the logical end of investigation (Schiappa, 2003, 115). In
this context, entitlement may be comprehended as producing the realities
of policy activity. In venturing to understand this phenomenon, the
notion of 'policy entitlement' outlined below is a multi-tier process by
which such a policy reality is discursively constructed, gains traction, is
advanced by parties to policy debates, and is ultimately adopted through
statutory policy provisions. For purposes of clarity, the description below
has been divided into a three-stage process comprising *interpretation,
resonance* and *modality*. However, these various processes may occur in

parallel or overlap as the new policy concept is propagated in different organisational, professional and/or geographic quarters.

Interpretation

Entitlement involves the naming of something. More specifically it may be conceived as the calling into existence through language that which of its own accord does not enjoy ontological status outside what is attributed to it by the naming process. In such circumstances, the named (entitled) entity is endowed an ontological status as, for example, an object, event, substance or vague feeling (Schiappa, 2003). It is through this status-endowing process that both the existence of something, and consequently our assumptions on how that thing can be known, are given (Burke, 1966). Thus, in calling something into existence through naming, entitlement furnishes a reference. It is from this reference that assumptions regarding existence status and ensuing knowledge deductions can be constructed.

However, such socially-coordinated 'epistemic access' (Boyd, 1993, 483) does not imply uniformity of interpretation regarding the properties of an entitled entity. Rather, a reference may only supply a shared path for interpretation (Ricoeur, 2002[1975]). The specifics of the interpretation produced are dependent on the subjective perspectives of the interpreter, albeit the latitude for subjectivity is constrained by the conventions of language use (Johnstone, 2018). This is most apparent when that which is entitled involves bestowing a unifying ontological status and consequent epistemological assumption on a set of entities conventionally considered separate and conceived differently. In such circumstances, as may be observed in the emergence of new policy concepts, the introduction of a new term may be required. However, given the need to reduce the complexity of ontologically unifying an assortment of entities normally considered discrete, the term employed would likely need to be a familiar word or conjunction of familiar words applied in a new context (Hart, 2008). This both dissipates potential rejection of the newly-entitled entity through the appearance of familiarity while concurrently directing interpretation of the entity's ontological and associated epistemological status. In this situation, appreciating the entitled entity involves an imaginative leap by the interpreter in transferring comprehensions of the familiar onto the entitled entity to reduce levels of abstraction. Thus, although the entitled entity is itself given independent ontological status, the entitlement process operates via use of metaphorical reasoning, wherein 'we talk and, potentially, think about something in terms of something else' (Semino, 2008, 1).

Thus, interpretation of the ontological status and associated epistemological position of the entitled entity is achieved by way of *connotative reasoning*. This understanding of something new and abstract in terms of the familiar prompts perceptions on the *apparent simplicity* of comprehending that which is entitled. Consequently, metaphors may provide particularly powerful vocabulary in which to perceive new policy concepts within the existing discursive field of established disciplines. However, in transferring connotations from the familiar onto the new or abstract in this manner, such reasoning not only enables appreciation of the introduced entity, but also constitutes how that entity is to be perceived. Furthermore, employing connotative reasoning facilitates a degree of polysemantic latitude in the transfer of attributes (connotations) from the familiar to the new entity. This latitude forms a reciprocal arrangement wherein the apparent simplicity in conceiving the entitled entity assists in its variable application. In this way, the entitled entity may subsume multiple entities normally considered discrete. Such *flexible signification* thereby operates in a relationship of reciprocity with *connotative reasoning* and *apparent simplicity* as a triad of 'naming effects'. A diagrammatic representation of the 'naming effects' engendered by entitlement is presented in Figure 7.1.

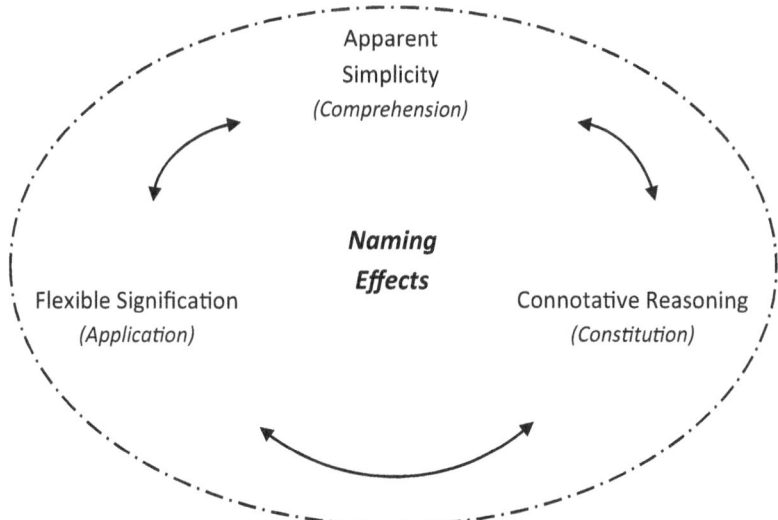

Figure 7.1 The *Naming Effects* of Entitlement
Source: author

Understood in the context of policy studies, this triad of naming effects facilitates the linguistic constitution of a concept (entitled entity) that may incorporate multiple problematic issues and thereby enable the provision of a unifying policy solution. For example, this was the case where a multitude of issues were subsumed beneath the label of 'green infrastructure' by means of *connotative reasoning* and *flexible signification* associated with the conjunction of the familiar words 'green' and 'infrastructure' (nature conservation, recreation provision, flood management, etc.). Concurrently, *apparent simplicity* and *connotative reasoning* facilitated deductions from the word 'infrastructure' so that such issues were both necessary for society and could be planned in much the same way as conventionally-conceived 'infrastructure'.

However, this mediation of meaning by *connotative reasoning* cannot be objective, as it obliges the interpreter to subjectively invest that which is being interpreted with a signification it does not already possess by way of existing formal denotation (Black, 1962; Knowles and Moon, 2006; Musolff, 2016). In this context, it can be conjectured that in the use of entitled concepts 'policy analysts are situated knowers thinking and writing from particular points of view' (Yanow, 1996, 27). Thus, the *flexible signification* stimulated by a concept's entitlement facilitates its appropriation for the needs of the end user.

As the entitled concept potentially encompasses a multitude of issues, this capacity for the concept to be 'positioned' relative to the requirements of the user may nonetheless result in recalibrating the relationship between that which is encompassed by the concept and the user of that concept. In this sense, the process of entitlement may be seen to do policy work by repositioning power relationships between the issues encompassed by the entitled concept and those who advocate the unifying policy solution deemed to address such issues. Through this process, resolution is thereby brought to problematic policy ambiguity by begetting a reality wherein those formulating policy have the power to define the source of the problems in question, and the ability to specify how they may be remedied.

Resonance

For a policy approach to gain traction among those parties concerned with the issues encompassed by the entitled concept, the principles upon which the policy approach is predicated must resonate with the prevailing rationalities to which those parties adhere. Such *practice accord* may manifest in a perceived ability to easily tailor existing policy formulation and implementation methods to the requirements of the new policy approach. Similarly, it may manifest in a perceived capacity to employ familiar language or disciplinary jargon in discussing the entitled concept when deducing from the 'naming effects' the appropriate interpretation of both the concept's meaning and applicability. For example, this was demonstrated with respect to the emergence and evolution of the GI planning approach in Ireland, where antecedent policy concepts and methods focused on the 'design', 'construction' and 'management' of 'infrastructure' and 'networks' was transferred onto new ideas for 'delivering' nature conservation via GI planning.

To achieve purchase, the perceived legitimacy to enunciate issues by those advocating the policy approach must be respected by those to whom it is introduced. This qualification of legitimacy, or *enunciative position*, is directly dependent on the rationalities to which the parties subscribe. For example, those advocating a GI planning approach in Ireland sought to harmonise with modernist rationalities in seeking legitimacy for their arguments. This entailed a process of 'stake inoculation' wherein effort was expended to represent GI as grounded in impartial and scientifically-sourced knowledge claims removed from subjective opinions. Here, symbolic acts of mapping, counting and comparison were used to produce an array of plans, diagrams, reports and references that weaved a web of signification which reinforced the enunciative position of those promoting GI.

The policy approach must also be perceived to possess a degree of *functional advantage* above that of existing policy approach(es) used in tackling the problematic issues subsumed by the entitled concept. A diagrammatic representation of the attributes necessary for the 'rationality resonance' of a new policy approach and the relationships between this and the 'naming effects' of entitlement is presented in Figure 7.2 below.

While the perceived *functional advantage* of a new policy may generally relate to its opined efficiency and effectiveness relative to that of previous policy or lack of policy, it is feasible that its gain may also lie, simultaneously or not, in the *enunciative position* it endows upon those introducing the concept. This possibility is resultant from the *flexible signification* engendered by a concept's entitlement which permits its interpretation, use and promotion in addressing the specific requirements of its advocate(s). In this way, the functions to which the entitled concept is put may be dependent on the intent of those advocating its introduction. Thus, it is conceivable that should a corollary of such advocated employment involve the reallocation and/or consolidation of the power to enunciate on an issue, an advantage of introducing an entitled concept may be perceived, tacitly or otherwise, as its (re)

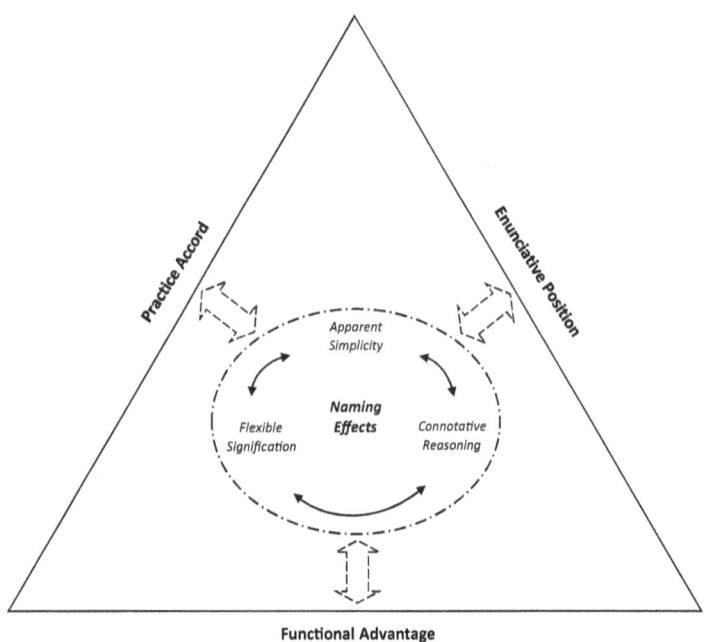

Figure 7.2 The Relationships between *Naming Effects* and *Rationality Resonance*
Source: author

distribution of enunciative legitimacy. As such, those advocating for the introduction of a new policy concept may do so in seeking to advance and/or reinforce the position of authority accorded to their profession.

Discourse Coalitions

Many of those who advance a discourse analysis approach to the study of policy propose that what bestows currency on ideas in the policy process is the common subscription to a specific narrative that clarifies meaning in situations of policy ambiguity (Epstein, 2008; Zahariadis, 2003; Martinez, 2021). Such 'discourse coalitions' permit the discernment of shared adherence to a particular narrative while simultaneously allowing a multitude of interpretations of the meaning embodied in that narrative. In this sense, the power to engender a discourse coalition may be conceived as deriving from its ability to facilitate 'discursive affinities' among the many issue-specific narratives of those parties involved. Hajer suggests that such affinities may not refer to actors and their intentions but rather 'operationalizes the influence of discursive formats on the construction of problems' (1995, 67).

This capacity of discourse affinities to shape the conception of the policy problems has been shown with respect to the evolution of GI discourses in Ireland. In particular, various discursive affinities associated with both the words 'green' and 'infrastructure' facilitated coalescence around GI narratives of numerous issues and interests heretofore considered separately. Those subscribing to the narratives of the GI discourse coalition did so consequent on the perceived advantages bestowed in reconceptualising their policy issues in terms of 'infrastructure'. From this perspective, agents could forge arguments regarding the importance of their specific issues of concern. Such reconceptualisation also offered the prospect of resolving problematic policy ambiguity by suggesting that longstanding issues of problem identification and decision ambivalence could be remedied through the logics employed in relation to conventionally-conceived 'grey' infrastructure. Thus, a perception emerged that nature conservation, recreational facilities provision and other complex policy issues could be solved via the traditional modes of planning, design, delivery and management associated with civil and services engineering.

In addition, the GI planning approach promoted concord between recreation planners, conservation ecologists and flood management engineers. Achieving this required the suspension of plausibly perceived incompatibilities or logical inconsistencies in providing a unifying policy solution to an assortment of potentially disparate problems. This was accomplished by means of the 'mythic' qualities of those narratives stemming from GI's entitlement.

Myth

The ability to suspend possible differences in advancing a narrative from which all parties to the discourse coalition are perceived to profit accords with the theoretical application of myth by Yanow in the context of policy and organisational analysis (Yanow, 1996, 2000). The term myth is employed here to designate a 'narrative created and believed by a group of people which diverts attention from a puzzling part of their reality' (Yanow, 1996, 191). The idea of myth forwarded in this context is not conceived as an evaluation of a narrative's veracity. In this sense,

It is not a question of whether a given description is an objective picture of reality but whether a given description receives the intersubjective assent of relevant members of a discourse community.

(Schiappa, 2003, 111)

As such, myth in terms of the analytical approach advanced here refers to a particular narrative format that facilitates subscription by a broad range of issue-specific interests through proffering apparent commensurability in situations where plausible discrepancies may coincide. Demonstrated in Chapter 5 is how through the process of *rationality resonance*, entitlement facilitates the naturalisation of a *mythic narrative* by its perception as the 'common sense' or 'natural' view of things. Shown here is how the reconceptualisation of green spaces as infrastructure was seen to 'make sense', and in doing so anthropocentrically repositions nature as that which could be planned to service society in much the same manner as conventionally-conceived infrastructure. The relationship of such a *mythic narrative* to the processes of entitlement is presented in Figure 7.3 below.

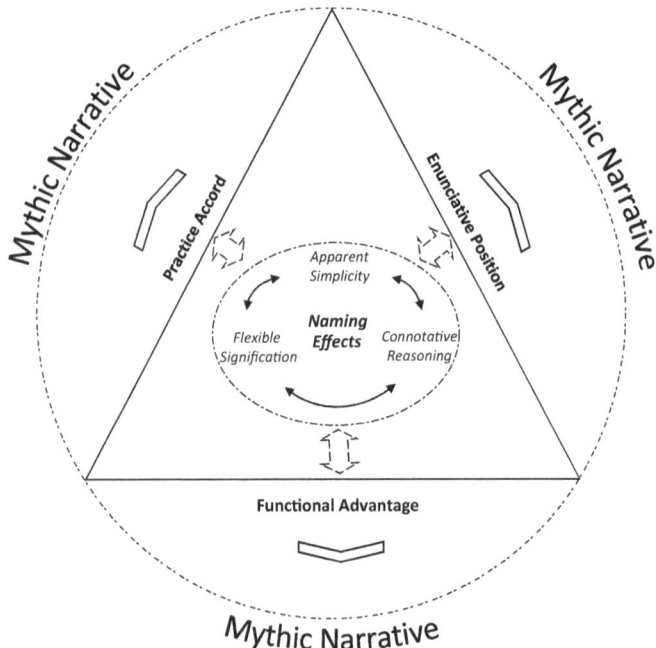

Figure 7.3 The Relationships between *Naming Effects, Rationality Resonance* and *Myth*

Source: author

Modality

By enabling the suspension of potential inconsistencies, the mythic qualities of the narrative prompted by concept entitlement facilitates the propagation of a policy approach through allowing multiple parties with various interests to espouse the same broad set of principles for problem resolution. In this way, myths do policy work by enabling those who subscribe to them to cooperate on issues in respect of which there is significant potential for disagreement. Accordingly, *mythic narratives* achieve what may be termed *modality* by capacitating the legitimate association of a policy approach with an array of problematic issues as well as existing problem–solution storylines. Expansion of the policy approach's discourse coalition is thereby facilitated.

Resultant from the *flexible signification* enabled by the 'naming effects' of concept entitlement, those seeking remedy to unsolved problems may adopt the *mythic narrative* as a resolving discourse. This phenomenon is here termed 'solution adoption'. Also, those desiring to communicate the merits of their existing policy activities may rebrand their current problem–solution storylines in a manner that harmonises with this narrative. This phenomenon is here termed 'elective rebranding'. Given the choice of those with problematic issues and problem–solution storylines to respectively adopt the narrative and electively rebrand their storyline in accordance with it, in such contexts, a process of *narrative volunteering* can be considered as operative.

Those employing the *mythic narrative* may now seek to consolidate support for their policy interests by assigning that narrative to other issues beyond their immediate concerns. This may be undertaken on the assumption that widening the applicability of issues which the policy approach is seen to address strengthens the perceived value of the policy approach they now subscribe to. Thus, advocates of this narrative may promote its coupling to other problematic issues which they perceive as lacking a coherent solution narrative. Additionally, such advocates may seek to rebrand as examples of the narrative's policy approach, existing problem–solution storylines circulating in other policy arenas. In such contexts, a process of *narrative application* can be deemed as operative. *Narrative application* thus operates via analogous but diametrically positioned processes to *narrative volunteering*. Accordingly, 'problem appropriation' and 'imposed rebranding' characterise this process. The relationship of such processes to those previously discussed is summarised in Figure 7.4.

The explanation advanced here holds that those advocating a mythic narrative may promote it as justifiably incorporating and representing both the problems and problem–solution storylines circulating on a broad

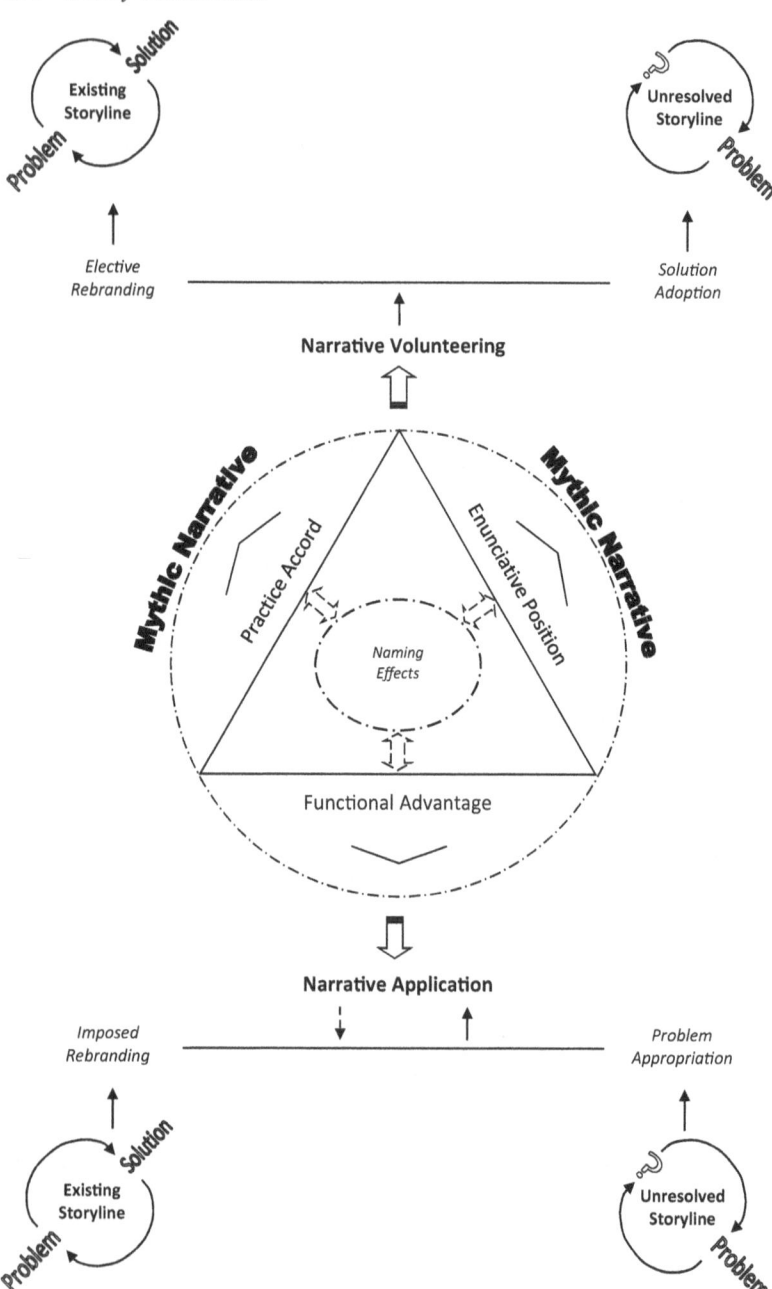

Figure 7.4 Diagrammatic Representation of *Policy Entitlement* in the Relationships between *Naming Effects, Rationality Resonance, Myth* and *Modality*
Source: author

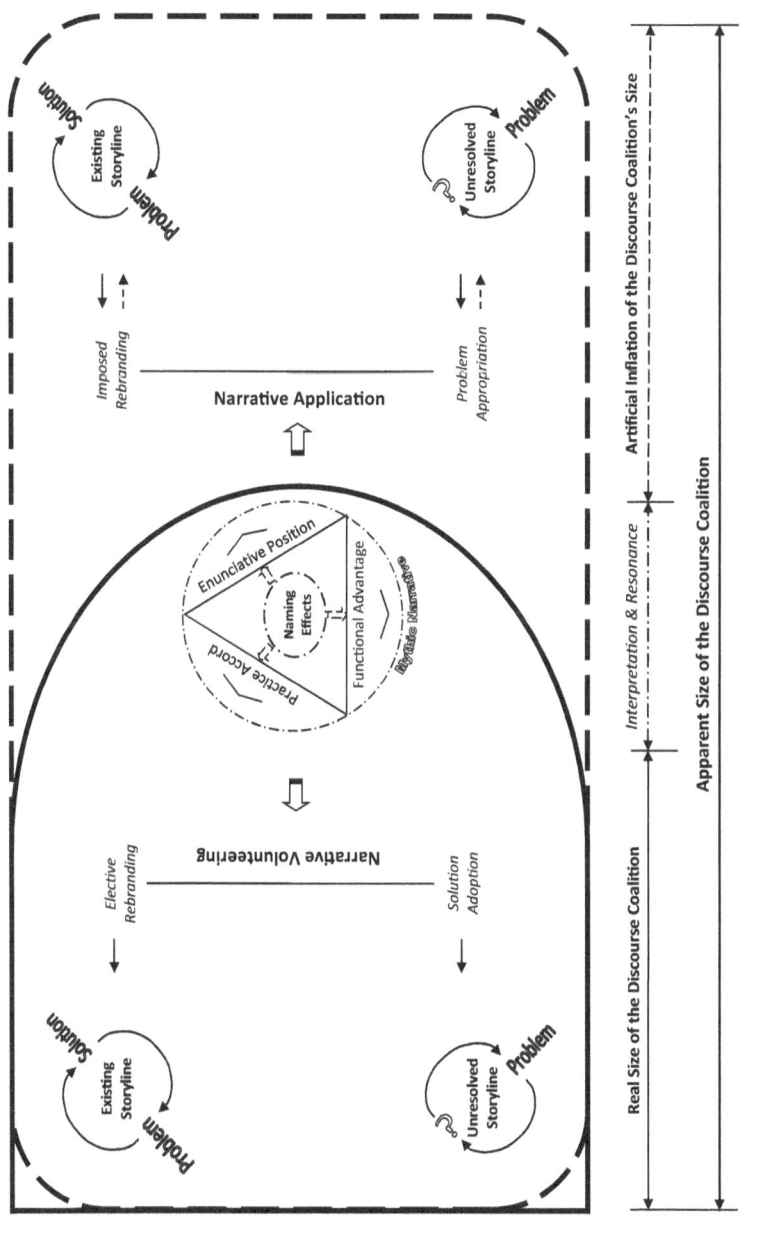

Figure 7.5 Artificial Inflation of Discourse Coalition Size by *Narrative Application*
Source: author

array of policy issues. This is achieved by means of *narrative volunteering* and *narrative application*. Consequently, the scope of issues addressed by the policy approach implied by the mythic narrative is perceived to expand with a corresponding increase in the apparent size of its discourse coalition. Thus, the conception of discourse coalition dynamics presented here allows for the artificial inflation of the coalition's apparent size through its member's endeavours to rebrand the storylines of others not necessarily supportive of the coalition's assertions. This may be undertaken as a deliberate strategy to advance a particular policy concept. As such, it extends the notion of discourse coalitions proposed by Hajer and others (Hajer and Laws, 2006; Epstein, 2008; Martinez, 2021), which advances a consensual process wherein 'actors group around specific story-lines that they employ' (Hajer, 1995, 13) [emphasis added]. Accordingly, this explanation adds depth to our appreciation of how such coalitions may form and function in policy advocacy. This process is summarised in Figure 7.5.

Enhancing Knowledge of the Policy Process

Benefits of an Interpretive Approach

The explanation of policy dynamics presented above conceives the policy process as an interrelationship between the dimensions of *signification, significance* and *applicability*. By giving prominence to the position of subjective and intersubjective interpretation, this approach avoids the determinism of policy process theory based on reductive rational choice models of social activity, Marxist-inspired explanations of ideological hegemony, or the inevitabilities characteristic of much institutionalist thinking. In attending to the detailed mechanics of interpretation, this explanation also evades the vagueness of explanations reliant on macro-sociological theories, such as structuration theories, to describe the traction of new policy concepts. Furthermore, the explanation of policy process dynamics presented here describes how a new policy concept may emerge, evolve and be adopted in the absence of dispute in situations where there exists significant potential for disagreement. As such, this explanation departs from the adversarialism presupposed by much policy process theory in both the positivist and post-positivist traditions.

Agent Positioning

The approach advanced here involves an appreciation of how those who participate in policy advocacy are 'positioned' by the discourses they promote relative to the subjects of deliberation, fellow advocates and other potential stakeholders. Thus, how agents align themselves with and shape different discourses may be understood to specify the power positions from which social actors can communicate and act with influence. In this way, discourses can be seen to part-constitute the identities of social actors by creating particular 'subject positions'.

Such positioning is facilitated by how governing activity in modern western democracies is intrinsically linked to perceptions of professional competence grounded in the possession of knowledge deemed credible by modernist rationalities. Consequently, those perceived as possessing this knowledge assume identities constituted by power relationships, and enjoy relative to others, the capacity to identify, control and legitimise the very issues taken to be the subjects of deliberation. This is facilitated through asymmetrics in the ability to deploy legitimating forms of knowledge that harmonise with prevailing empiricist rationalities. This phenomenon is demonstrated in Chapter 5, where it is outlined how endeavours to represent GI as rooted in scientific methodologies influenced the form of its advocacy, perceptions of its meaning and the credibility of those who enunciated upon it. Detailed is how practices of 'stake inoculation', such as cartography and quantification, were employed to present the impression of impartiality seen as crucial to resonating with the 'technical-rational model' of knowledge production prevalent in planning practice.

Hence, forging subject positions through presentation constitutes an important element of governing activity. This is consequent on the contention that 'the question of who should have the authority to make definitional decisions amounts literally to who has the power to delineate what counts as Real' (Schiappa, 2003, 178). Thus, what is asserted here accords with a Foucauldian concept of power relations, in which different forms of knowledge in different contexts result in the allocation of power to those who can deploy such knowledge through the perceived legitimacy of their enunciations (Foucault, 1980). As such, it presents an understanding of power as dispersed throughout numerous sites in the policy process rather than singularly located.[1] For example, in the case of GI's ascension in Ireland, this is demonstrated by the influence and activities of those who were seen as able to deploy the apparent impartiality of 'objective' scientific methods of knowledge production. In this way, meaning making in and through the policy process may function to enhance the positions of certain individuals, organisations and/or networks through engendering a knowledge–power–identity nexus perceived as possessing the capacity to deploy means of knowledge production that accords with prevailing rationalities.

Conclusion

Attention to the role of meaning making moves beyond conventional assumptions of the policy process as a form of 'applied problem-solving' wherein policy formulation is conceived as a linear progression from problem identification to solution specification. Accordingly,

> This step to treat policy practice as the site at which interpretive sche-
> mata are produced and reproduced is a significant one. It builds on the
> linguistic account of policy making that employs narratives – stories,
> metaphors, myths – to create an image of the world that is acted upon
> and that constitutes that world at the same time.
>
> (Hajer and Laws, 2006, 264)

This challenges much policy process theory, which erroneously partitions the 'real' and the 'representational'. This is achieved by demonstrating how '[p]olicies and political actions are not either symbolic or substantive. They can be both at once' (Yanow, 1996, 12). The idea of 'policy entitlement' proposes one way of identifying and explaining the 'interpretive schemata' that give traction to new concepts within policy formulation activity. Fundamental to this explanation is a focus on the role of symbolic language, acts and objects in both constituting and communicating interpretations of a new policy concept's signification, significance and applicability. Central to this explanation is that meaning making both in and through policy is a contextual process wherein prevailing professional rationalities determine what a policy is perceived to remedy and how it is advocated. Consequently, this explanation addresses lacunae in our understanding of how the policy process may engender new perceptions of reality that render problematic policy ambiguity amenable to resolution by existing policy formulation practices. This is achieved by outlining how the interpretation of meaning relative to prevailing rationalities may 'will to truth' (Foucault, 1976, 55) the realities that enable problem definition and make solution specification possible. Therefore, this explanation offers a method to explore how bringing 'new possibilities into being is, of necessity, to introduce new criteria for the objective application of the new ideas that permeate our world' (Hacking, 2002, 23). The explanation forwarded here thereby opposes the view prominent in much policy process theory, such as the Advocacy Coalition Framework (Weible and Sabatier, 2018), by emphasising that '[m]eanings are not just representations of people's beliefs and sentiments about political phenomena; they fashion these phenomena' (Wagenaar, 2011, 3).

This explanation also addresses deficits in our knowledge of how agents resolve issues of problematic policy ambiguity in a manner that facilitates seemingly unanimous support for a policy proposal where there exists significant potential for dispute. This is accomplished by showing how the interpretive requirements of ambiguous signification facilitate a degree of polysemy tempered by the prevailing rationalities of professional practice. In stressing the role of *mythic narratives* in suspending logical inconsistencies and deflecting potential criticism, this explanation demonstrates the function played by such ambiguous signification in furnishing the means through which various motivations may be synchronised to make action possible.

Accordingly, what the idea of 'policy entitlement' proposes is that in cases where problematic policy ambiguity appears resolved by common subscription to a new concept, what the analyst should attend is how symbolic language, acts and objects suspend potential conflict between different interests. It is here that access may be gained to the inter-pretive schemata of contextually-situated policy practice. Through this, the rationalities underpinning such practice(s) and their potential impacts may be revealed.

Note

1 The concept of power advanced here is one delimited by a requirement to resonate with prevailing rationalities. Thus, the focus is on how legitimate enunciation is explicitly and implicitly both restricted and enabled, as well as how its source generates particular effects. Consequently, this conception of power differs from Lukes' understanding of power dynamics as surreptitious 'power over' others (Lukes, 2005).

References

Black, M. 1962. *Models and Metaphors*. Ithaca, New York, U.S.A., Cornell University Press.

Boyd, R. 1993. Metaphor and theory change: what is 'metaphor' a metaphor for? In: Ortony, A. (ed.) *Metaphor and Thought*. Cambridge, England, U.K., Cambridge University Press.

Buchstein, H. and Jorke, D. 2012. The argumentative turn toward deliberative democracy: Habermas's contribution and the Foucauldian critique. In: Fischer, F. and Gottweis, H. (eds.) *The Argumentative Turn Revisited: public policy as communicative practice*. London, England, U.K., Duke University Press.

Burke, K. 1966. *Language as Symbolic Action*. Berkley, California, U.S.A., University of California Press.

Dryzek, J. S. 2012. *Foundations and Frontiers of Deliberative Governance*. Oxford, England, U.K., Oxford University Press.

Epstein, C. 2008. *The Power of Words in International Relations: Birth of an Anti-Whaling Discourse.* London, England, U.K., The MIT Press.

Feldman, M. S. 1989. *Order Without Design: Information Production and Policy Making.* Stanford, California, U.S.A., Stanford University Press.

Fischer, F. 2003. *Reframing Public Policy: Discursive Politics and Deliberative Practices.* Oxford, England, U.K., Oxford University Press.

Fischer, F. and Gottweis, H. (eds.) 2012. *The Argumentative Turn Revisited: Public Policy as Communicative Practice.* London, England, U.K., Duke University Press.

Foucault, M. 1976. *The Will to Knowledge: The History of Sexuality: Volume 1.* London, England, U.K., Penguin Books Ltd.

Foucault, M. 1980. *Power/Knowledge: Selected Interviews and Other Writings 1972–1977.* New York City, New York, U.S.A., Pantheon.

Goodin, R. E., Rein, M. and Moran, M. 2006. The public and its policies. In: Goodin, R. E., Rein, M. and Moran, M. (eds.) *The Oxford Handbook of Public Policy.* Oxford, England, U.K., Oxford University Press.

Hacking, I. 2002. *Historical Ontology.* Cambridge, Massachusetts, U.S.A., Harvard University Press.

Hajer, M. 1995. *The Politics of Environmental Discourse: Ecological Modernisation and the Policy Process.* Oxford, England, U.K., Oxford University Press.

Hajer, M. and Laws, D. 2006. Ordering through discourse. In: Moran, M., Rein, M. and Goodin, R. E. (eds.) *The Oxford Handbook of Public Policy.* Oxford, England, U.K., Oxford University Press.

Hart, C. 2008. Critical discourse analysis and metaphor: toward a theoretical framework. *Critcial Discourse Studies,* 5, 91–106.

Healey, P. 2005. *Collaborative Planning: shaping places in fragmented societies.* Basingstoke, England, U.K., Palgrave Macmillan.

Healey, P. 2012. Performing place governance collaboratively: planning as a communicative process. In: Fischer, F. & Gottweis, H. (eds.) *The Argumentative Turn Revisited: Public Policy as Communicative Practice.* London, England, U.K., Duke University Press.

Hill, M. and Varone, F. 2021. *The Public Policy Process.* Abingdon, England, U.K., Routledge.

Johnstone, B. 2018. *Discourse Analysis.* Hoboken, New Jersey, U.S.A., Wiley.

Kingdon, J. W. 1984. *Agendas, alternatives and public politics.* New York City, New York, U.S.A., Harper Collins.

Knowles, M. and Moon, R. 2006. *Introducing Metaphor.* Oxford, England, U.K., Routledge.

Lukes, S. 2005. *Power: A Radical View.* Basingstoke, England, U.K., Palgrave MacMillan.

Martinez, M. 2021. Discourse coalitions and the messiness of policy solutions: college governance in Nevada. In: Zittoun, P., Fischer, F. and Zahariadis, N. (eds.) *The Political Formulation of Policy Solutions.* Bristol, England, U.K., Bristol University Press.

Musolff, A. 2016. *Political Metaphor Analysis: Discourse and Scenarios.* London, England, U.K., Bloomsbury Publishing.

Myerson, G. and Rydin, Y. 1996. *The Language of Environment: A New Rhetoric.* London, England, U.K., UCL Press.

Rhodes, R. A. W. 2018. *Narrative Policy Analysis: Cases in Decentred Policy.* London, England, U.K., Palgrave Macmillan.

Ricoeur, P. 2002[1975]. *The Rule of Metaphor.* Oxford, England, U.K., Routledge.

Schiappa, E. 2003. *Defining Reality.* Edwardsville, Ilinois, U.S.A., Southern Illinois University Press.

Semino, E. 2008. *Metaphor in Discourse.* Cambridge, England, U.K., Cambridge University Press.

Stone, D. 2012. *Policy Paradox: The Art of Political Decision Making.* New York City, New York, U.S.A., W.W. Norton.

Wagenaar, H. 2011. *Meaning in Action: Interpretation and Dialogue in Policy Analysis.* Armonk, New York, U.S.A., M.E. Sharpe.

Weible, C. M. and Sabatier, P. A. 2018. *Theories of the Policy Process.* Abingdon, England, U.K., Routledge.

Yanow, D. 1996. *How Does a Policy Mean? Interpreting Policy and Organisational Actions.* Washington DC, U.S.A., Georgetown University Press.

Yanow, D. 2000. *Conducting Interpretive Policy Analysis.* London, England, U.K., Sage Publications.

Yanow, D. 2006. Accessing local knowledge. In: Hajer, M. and Wagenaar, H. (eds.) *Deliberative Policy Analysis: Understanding Governance in the Network Society.* Cambridge, England, U.K., Cambridge University Press.

Yanow, D. 2007. Qualitative-interpretive methods in policy research. In: Fischer, F., Miller, J. G. and Sidney, M. S. (eds.) *Handbook of Public Policy Analysis: Theory, Politics and Methods.* Boca Raton, Florida, U.S.A., CRC Press.

Zahariadis, N. 2003. *Ambiguity and Choice in Public Policy: Political Decision Making in Modern Democracies.* Washington DC, U.S.A., Georgetown Press.

Index

agent positioning 138 *see also* enunciative
ambiguity 1–2, 130
ambiguous signification 62
apparent simplicity 72–75, 129–130
associative interpretation 62–63

balance 110–110
Barthes, R. 21–22, 27, 62, 64, 71, 78n1, 108
Berger, P. 20, 63, 72
biodiversity 39, 42, 47–48, 85, 91, 92, 97–99, 111
Boyd, R. 22, 73–74
broader problem narrative 49–50
Burke, K. 20–21, 59–60, 127–128

categories 20
theories of knowledge: 59, 61
Comhar 43, 48–50, 53, 56n1, 57n2, 71, 89, 91, 97, 118–119
comparison 23
constrained idealist 5
conceptual metaphor 67–68
connotative reasoning 71–72, 129–130
counting 22–23: *see also* stake inoculation by quantification

deflection 110–110
discourse analysis 12–17
discourse coalitions 17–19, 132–133

discursive: affinities 18, 105–107, 112, 132–133; contamination 18, 105–107, 112; weight 52–54
Dublin City Council 85, 92

ecological network 42, 47
elective rebranding 115–117, 135
epistemological 21–22, 25, 60–61, 73, 128–129
Epstein, C. 15, 19
entitlement 21, 59–60 *see also* policy entitlement
enunciative: advantage 88; position 131
Environmental Protection Agency (EPA) 47
existing planning vehicles 87–88

Feindt, P. 13
Fingal County Council 40
flexible signification 75–77, 129–131, 135
footing *see* stake inoculation
Foucault, M. 15–16, 96
functional: expectations 70–71; advantage 98–101, 131

Galway City Council 43, 85
Goffman, E. 24, 93
governmentality 4
Green City Guidelines 42, 47–48
green infrastructure: description 38–39; Ireland 41–43

Hajer, M. 18–18, 22, 105, 112, 138, 140

imposed rebranding 119–121, 135
interpretation 128–130
interpretive policy analysis 2–5, 125–127

Johnson, M. 67–68

Kingdon, J.W. 2, 14, 20, 22, 23, 96
knowledge: for policy 11; of the policy process 11–12
Knowles, M. 64, 68

Lakoff, G. 67–68
language familiarity 84–87
Luckmann, T. 20, 63, 72

maps 25–26
meaning making *see* interpretive policy analysis
Medina, J. 59, 74–75
metaphor 21–22, 64–70, 73, 129
modality 135–138
Moon, R. 64, 68
Myerson, G. 21, 22, 26, 126
myth 26–27, 107–108, 132–135

naming attributes 61–64
naming effects 64–71, 129–130
narrative: application 113, 117–121, 135–138; of necessity 54–55; modality 111–113; volunteering 112–117, 135–138
naturalisation 62, 64, 71–72, 75, 134

Oels, A. 13
ontological 21–22, 25, 60, 67–68, 77–78, 127–129
ontology 78

persuasion 2
policy approach 39
policy entitlement 127–128

policy persuasion *see* persuasion
Potter, J. 22, 24, 93, 98
post-political 4
practice accord 83–88, 130–131
problematising 47–50
problem: ambiguity *see* ambiguity; appropriation 117–118, 135

rationality resonance *see* resonance
recreation 92
recurrence 69–70
repetition 68–69
resonance 130–132, 134
Ricoeur, P. 66–67, 76, 78
root problem narrative 47–48
Rydin, Y. 21, 22, 26, 126

Semino, E. 68, 73–74
Schiappa, E. 12, 21–22, 60, 72, 127–128, 134
Schön, D.A. 21, 54, 56, 111–112, 123n2
social constructionism 5
solution adoption 113–115, 135
solving 50–54
stake inoculation 24, 131: by comparison 93–95; by mapping 89–93; by quantification 95–98
Stone, D. 16, 19, 23, 96
storylines 17
subject positions 15–16, 139
suspension 107–110
sustainable development council *see* Comhar
sustainable urban drainage systems (SUDS) 99, 109, 123n1
symbolic: acts 22–24; language 19–22; objects 24–26

Torfing, J. 13
transport 86, 114

Yanow, D. 19, 22, 26, 76, 107–108, 126, 130, 133

Zahariadis, N. 2